Practical Guide to Managing Acidic Surface Waters and Their Fisheries

Robert W. Brocksen
Executive Director
Living Lakes, Inc.
Greenbelt, Maryland

Michael D. Marcus
President
Western Aquatics, Inc.
Laramie, Wyoming

Harvey Olem
Owner, Olem Associates
President, The Terrene Institute, Inc.
Washington, D.C.

LEWIS PUBLISHERS
Boca Raton   Ann Arbor   London   Tokyo

Library of Congress Cataloging-in-Publication Data

Catalog record is available from the Library of Congress.

International Standard Book Number 0-87371-755-4

LEWIS PUBLISHERS, INC.
121 South Main Street, P.O. Drawer 519, Chelsea, Michigan 48118

Printed in the United States of America   1 2 3 4 5 6 7 8 9 0

ACKNOWLEDGMENTS

Many individuals have contributed to this book in a myriad of ways. We thank Living Lakes, Inc., the Electric Power Research Institute, and the United States Fish and Wildlife Service for providing funding for this project. While it is impossible to credit all individuals who have helped in the preparation of the book, we wish to acknowledge the contributions of the following: Timothy B. Adams, Sally A. Brocksen, Leslie B. DeMarco, Walter E. Eifert, Gretchen Flock, David Gulley, William B. Harrison III, Jaye Douglas Isham, Carrie Marchese, Rebecca A. Nichols, Donald Porcella, Lynne Poslethwaite, R. Kent Schreiber, Lucille Shandloff, Carol R. Shriner, Lura Svestka, and Judith Taggart.

PREFACE

This book resulted from our desire to present in one volume important practical information gathered over the past several years on management of acidic surface waters and their fisheries.

Unlike the Scandinavian countries whose liming programs are government-sponsored, North America has no federally funded program to manage acidic waters. Therefore, this book has melded many resources — both federal and private — and in addition, related this information to guidance manuals prepared by our government colleagues in Norway and Sweden.

The book complements another Lewis publication, *Liming Acidic Surface Waters*. Published in 1991, this liming book is a comprehensive integration of European and North American research that provides readers with a thorough understanding of the complexities involved with liming.

The *Practical Guide to Managing Acidic Surface Waters and Their Fisheries* gives readers the basic tools to perform the necessary steps involved in managing acidic surface waters. It summarizes the scientific and technical information evaluated in the liming book in a practical manner and provides the information needed to carry out a variety of acidic surface water management activities.

We hope the *Practical Guide to Managing Acidic Surface Waters and Their Fisheries* will be used by fishing club managers, state and local government personnel, lake association managers, and others responsible for maintaining healthy lakes and streams that may be affected by acidity from any source.

R.W. Brocksen, M.D. Marcus, H. Olem

Robert W. Brocksen is Executive Director of Living Lakes, Inc., a not-for-profit organization financed by investor-owned electric utilities and coal producers and charged with the responsibility to demonstrate the efficacy of using limestone to neutralize acidified lakes and streams. Dr. Brocksen is an adjunct professor in the Department of Zoology and Physiology at the University of Wyoming. He received his Ph.D. in fisheries, physiology, and limnology from Oregon State University.

Michael D. Marcus is President and co-founder of Western Aquatics, Inc., in Laramie, Wyoming. The firm provides scientific expertise on fishery biology, water quality, aquatic habitat quality, aquatic toxicology, and water basin planning. For over 20 years, his research has emphasized relationships of habitat variables to ecological populations and communities in lakes, streams, and reservoirs. Dr. Marcus has a Ph.D. in zoology, with an emphasis on aquatic ecology, from the University of Wyoming.

Harvey Olem is owner and founder of Olem Associates, an environmental consulting firm in Washington, D.C., that provides decisionmakers with scientific assessments and research to improve protection of the environment. Dr. Olem is also President of Terrene Institute, a not-for-profit corporation that conducts research and educational activities related to the environment. He received his Ph.D. in environmental engineering from The Pennsylvania State University and is a registered professional engineer in four states.

CONTENTS

LIST OF TABLES

LIST OF FIGURES

CHAPTER 1

Introduction

The problems of acidity in surface waters and their effects on biological organisms, including fish, have been recognized for centuries. Until the 1970s, however, attempts to control acidity in surface waters for the management of fisheries were limited in scope. During that decade, however, international attention focused on atmospheric acidity and its resultant acidic deposition, or acid rain. This interest stimulated a huge volume of research worldwide to attempt to identify sources of the high levels of atmospheric acid, the mechanisms of its formation and transport, its effects on the total environment, including surface waters, and finally, mitigation of the effects of acid rain. This guide addresses how to mitigate those effects on fisheries through the application of lime, which improves the quality of surface waters and increases fish productivity.

The value of liming to fish culture includes five important benefits. First, it neutralizes and buffers acidity in both the water and sediment, improving environmental quality for fish and the organisms they feed on. Second, calcium added by liming can displace fertilizing substances from sediments to the water column, such as potassium and phosphate, making them more readily available. Third, the calcium carbonate in lime-stone can flocculate fine humus particles in the sediments, causing the mean sizes of soil particles to increase, thereby increasing sediment permeability and aeration. Fourth, dissolution of limestone can increase dissolved concentrations of carbon dioxide, enhancing the production rates of plants that are important sources of food and cover for fish. And fifth, in some extremely softwater ponds, liming can provide necessary metabolic calcium in concentrations that support satisfactory fish growth.

Liming, as used in this manual, is the addition of alkaline materials to surface waters or their watersheds for the purpose of mitigating, or neutralizing, acidic conditions. Limestone is the material most commonly used to reduce acid levels in waterbodies. Liming is not a new technique; it has long been used in agriculture and aquaculture to improve water and soil quality.

Much recent emphasis on liming lakes and streams stems particularly from concerns about mitigating the effects of acid deposition, particularly in areas of northern Europe and eastern North America. Sweden has the most extensive liming program among the Scandinavian countries. In 1977, the Swedish Government began an experimental program that now has limed more than 6,000 previously acidified lakes and rivers. The program emphasizes (1) improving waters of special value for fishery, nature conservation, or recreational use; (2) studying the ecological effects of liming; (3) evaluating alternative liming methods; and (4) estimating costs associated with liming. All waters with a pH of less than 6.0 and/or an alkalinity of less than 2.5 mg/L as $CaCO_3$ (50 µeq/L) are eligible for governmental subsidies for liming. The principal goal in Swedish liming of freshwater is to increase the pH to above 6.0 and the alkalinity to above 5 mg/L (100 µeq/L), and to detoxify the water so that naturally occurring flora and fauna can persist or recolonize the limed water.

Continued research on lake and stream liming in the United States has been sponsored principally by three organizations: the U.S. Fish and Wildlife Service (FWS), the Electric Power Research Institute (EPRI), and Living Lakes,

Inc. (LLI) (Olem et al. 1991). Such research is needed because, despite the fact that liming has been conducted extensively in Scandinavia and to a lesser extent in the United States and Canada, insufficient information exists (1) to predict site-specific responses by the many biological components in limed lakes and streams, and (2) to develop and implement a management level liming program. Excellent and detailed discussions of the history of liming, individual international liming efforts, and aspects of liming in general can be found in the literature listed in the reference section.

The importance of liming culture ponds to increase fish productivity was increasingly recognized by the early 1900s, particularly in Europe, for ponds built on soils having low calcium content. For example, a 1925 report notes that liming used alone increased the fish productivity in a single season in one European pond from 80 to 360 kilograms per hectare. Liming has been widely recommended since the late 1940s as a method to increase fish production in softwater culture ponds in the United States, particularly in the Southeast, where increases in fish productivity from 25 to 100 percent are common in limed ponds.

It is important to emphasize that surface water liming is not a form of fertilization; it is a remedial action to improve the habitat for fish. However, liming is often used in conjunction with fertilization to improve the productivity of fish ponds. For example, half of 10 fertilized fish culture ponds in Alabama were limed and productivity changes monitored for fish and fish food organisms. In contrast to the unlimed ponds, productivity by food organisms in the limed ponds increased by 128 percent, while fish increased by 125 percent.

While liming lakes and streams can increase the survival and growth of resident biota in these systems, aluminum and other metals can continue to leach from the surrounding watershed and adversely affect the aquatic biota. To reduce these effects, strategic liming of shorelines and wetlands is used to reduce metal leaching and prolong by several years the retention of alkalinity levels in lakes, even in lakes with relatively short retention times.

Soil liming of a watershed area can produce a variety of chemical, biological, and physical benefits. Direct chemical changes associated with liming acidic soils include decreases in hydrogen ion concentrations, as well as increases in calcium concentrations, plus percent base saturation. Magnesium concentrations also increase, especially when dolomitic limestone is used. Subsequently, these chemical changes can cause increased availability of some nutrients, especially phosphates, and decreased availability of metal, including iron, aluminum, and manganese. Liming can help prevent some metals, such as aluminum, from accumulating to potentially toxic levels in the soils.

In southwest Scotland, a research project experimentally limed the watershed surrounding Loch Fleet. While this lake originally supported a brown trout fishery, the fishery had disappeared about 20 years ago, apparently as a result of acidification. This watershed was first limed in the spring of 1986. After treating the waters draining the surrounding watershed, the lake itself, and the waters draining the lake, pH levels and calcium concentrations increased, while the aluminum concentrations decreased, especially the toxic, labile monomeric form. In the spring of 1987, the lake was stocked with brown trout, which had high survival rates, apparently permitted by the liming. Also, they successfully spawned during the following winter.

Liming waterbodies serves numerous environmental purposes and goals. In the short term, it preserves and improves the habitats of desirable aquatic species. In the long term, with the rapid acceleration of food fish depletion rates appearing regularly in news headlines, it can be a factor in preserving a basic source of the world's food.

The main objective of this guide is to aid waterbody managers in the use of neutralizing limestone materials to achieve and maintain suitable water quality for fisheries. The chapters that follow discuss the steps needed to determine if acidity is adversely affecting a fishery. They define the types of waterbodies that are good candidates for liming and suggest the best times to lime. In addition, if the decision is to lime, they provide the reader with the information and the processes necessary to obtain required permits from local, state, and federal agencies.

This guide also describes how to apply the limestone cost effectively for a particular body of water and what results to expect following the liming. It includes how to evaluate the success of liming and how to continue managing the surface waters through repeat applications of neutralizing materials. In summary, this guide will help the reader decide whether or not to lime, and, if the decision is to lime, it will help the reader and user through the steps necessary to carry out a successful application.

CHAPTER 2

Establishing Needs and Feasibilities in Surface Water Liming

Improving angling success is the primary motivation for most fishery enhancement efforts. Declines in angler catch, whether actual or perceived, coupled with observed surface-water acidity can suggest liming as a useful management tool. However, two questions should be answered before liming any surface water. Can liming help restore or protect the fishery? Is liming a feasible management approach for the water? This chapter introduces necessary considerations and useful approaches for analyzing the data to answer these two questions.

Before discussing specific criteria for determining whether to lime, we briefly overview the major water quality changes in surface water acidification and review how these changes can directly affect fish. This information introduces the scientific basis for understanding how liming can help improve water quality for fisheries and why it can be a potentially useful tool for fishery management.

WATER QUALITY CHANGES ACCOMPANYING SURFACE WATER ACIDIFICATION

Surface water acidification refers to increasing acidity in a lake or stream shown by decreasing pH levels. Acidification can occur because of acid input to water from various sources, including the atmosphere (acidic precipitation or "acid rain"), watershed (e.g., industrial discharges or acid mine drainages), or internal (microbial and chemical) processes. However, it is important to recognize that other chemical changes in water quality, in addition to increased hydrogen ion (H^+) concentrations, accompany acidification. Of these, the most easily measured is the buffering capacity of the water to resist acidity changes, which is routinely determined by measuring the alkalinity of the water.

In simple terms, alkalinity is a measure of the water's capacity to neutralize acids through dissolved concentrations of carbonate, bicarbonate, and hydroxide ions (respectively, CO_3^{2-}, HCO_3^-, OH^-). While these dissolved anions are primarily associated with dissolved calcium in most waters, they also can be associated with sodium, magnesium, and other metallic cations. The total acid neutralizing capacity (ANC) of waters depends not only on this chemical system directly but also can involve (1) organic acids and other anions that can buffer against increased acidity; (2) various microbiological processes occurring primarily in the sediments that consume acid-forming ions; and (3) photosynthesis and respiration rates by resident plants and animals that consume and release carbon dioxide (CO_2) in the water (Wetzel, 1983; Kelly et al. 1987).

Increased acidity also increases the solubility of many potentially toxic metals and the prevalence of their more toxic chemical ionic forms. Therefore, in nature increasing acidity accompanies both decreasing alkalinity and increasing concentrations of dissolved metals. Of these metals, aluminum especially tends to leach into surface waters from surrounding watersheds where acidic waters percolate through soils (Driscoll et al. 1988).

HOW ACIDIC WATERS AFFECT FISH

Field and laboratory studies confirm that three water quality variables—decreasing pH, low calcium concentrations, and elevated aluminum concentrations—regulate the effects of acidification on fish. The exact threshold and magnitude of these effects vary with species (Fig. 2-1). Some species begin to show stress from elevated hydrogen ion concentrations (i.e., increased acidity) when pH levels decline below 6.5; for most species, the critical level is between pH 6.0 and 5.0 (Marcus et al. 1986; Baker and Christensen, 1991). Toxic stress also can occur as concentrations of inorganic monomeric aluminum (including Al^{3+} and various associations of this cation with OH^-) increase above about 0.005 mg/L. Finally, calcium concentrations less than 2.0 mg/L as calcium can produce physiological stress in many fish and other aquatic species (Brown, 1982).

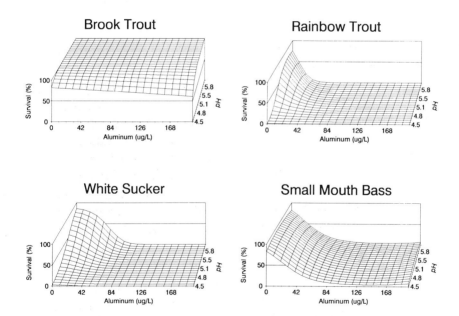

Figure 2-1.—Tolerance for swim-fry of four fish species to ranges of pH and aluminum concentrations in solutions of 2.0 mg/L calcium (data provided by Dr. Harold Bergman, Lake Acidification and Fisheries Project, University of Wyoming).

Often, one of the first and most apparent responses of fish to acidification is the absence of one or more entire age classes within some fish populations (Mills et al. 1987). These missing age classes result from a failure to recruit new year classes into a population. Their absence usually becomes apparent when sampling fish in lakes and streams to examine population age structure or size classes. For example, examination of a fish population sampled in late 1991 might reveal fish that were recruited to the population during 1984, 1985, 1987, and 1990 and that the year classes from 1986, 1988, 1989, and 1991 are missing. While such patterns are not unusual for some wild fish populations not stressed by acidity (e.g., white suckers in a variety of water qualities), these patterns can become increasingly common for fish species stressed during surface water acidification.

Results from some field studies indicate that a possible reason for recruitment failures is the reabsorption of eggs by female fish, causing spawning failures. Laboratory studies using a variety of fish species, however, fail to confirm this as a general mechanism that causes missing year classes for fish inhabiting acidic waters. Instead, these studies show that mature fish living in acidic water can have reduced feeding, loss of weight, and fewer eggs per female fish, while the average number of eggs per weight of female fish remains unchanged (Mount et al. 1988a,b). The laboratory results also indicate that these recruitment failures may be caused primarily by higher sensitivities of young fish to acidic conditions. For example, studies with brook trout show that freshly fertilized eggs, hatching fry, and swim-up fry generally have the greatest sensitivities to acid and aluminum toxicities (Fig. 2-2).

Although little information exists on the physiological mechanisms causing the mortality of young fish, increased dissolved calcium concentrations can clearly help protect both early and later life stages of fish from toxic effects of acidity and aluminum (Brown, 1982). In some acid-sensitive lakes and streams, calcium concentrations can be 1.0 mg/L or less. In comparison, laboratory studies show that calcium concentrations less than 4.0 mg/L can be stressful to many fish species, and concentrations less than 2.0 mg/L are stressful to most fish

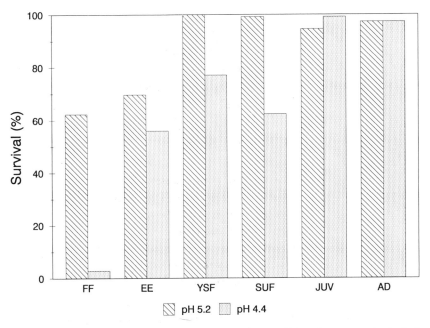

Figure 2-2.—Model predictions of effects of two pH levels on survival of six brook trout life stages inhabiting waters with 5 μg/L aluminum and 2 mg/L calcium (FF = freshly fertilized eggs, EE = eyed eggs, YSF = yolk-sac fry, SUF = swim-up fry, JUV = juvenile [fingerling] fish, AD = adults; from Mount and Marcus, 1989).

species (Marcus et al. 1986). Laboratory studies also show that fingerling and older fish that inhabit waters with elevated acid and aluminum levels and low calcium concentrations can exhibit (1) severe disruption of internal body-ion balances because of pH and/or aluminum effects; (2) severe respiratory stress, primarily the result of aluminum effects; or (3) both (Morris et al. 1988). Disruption of internal ion balances upsets various essential life functions, and respiratory stress leads to many symptoms, including suffocation. Either response can be lethal. While the extent of actual effects depends on the fish species and the acid, aluminum, and calcium in solution, increasing the dissolved calcium concentrations to 4.0 mg/L and greater clearly helps alleviate many stresses in fish (Fig. 2-3).

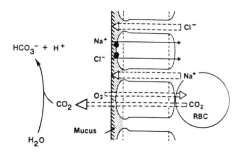

A. Structure and Function of Unstressed Fish Gill

- mucus coats external gill surface at water surface
- chloride cells "pump" Na^+ and Cl^- from water into fish
- Na^+ and Cl^- leak from fish between cell junctions to water
- red blood cells (RBC) exchange CO_2 for O_2 inside gill
- O_2 and CO_2 diffuse across epithelial cells of gill
- CO_2 reacts with water to form HCO_3^- and H^+ outside of gill

B. Decreasing pH (Increasing Acidity)

- displaces Ca^{2+} from gill surface
- leads to increased gill damage
- increases rates of Na^+ and Cl^- loss
- inhibits active Na^+ and Cl^- "pumping"
- causes ionoregulatory toxicity

C. Increasing Aluminum at Low pH

- aluminum binds to gill surface
- further displaces Ca^{2+} from gill surface
- leads to additional gill damage
- further increases Na^+ and Cl^- loss
- further inhibits active Na^+ and Cl^- uptake
- increases potential for ionoregulatory toxicity

D. Increasing Aluminum at Moderate pH

- aluminum precipitates on gill surface
- increases mucus production and inflammation
- increases diffusions distances for O_2 and CO_2
- reduces O_2 and CO_2 exchange rates
- increases potential for respiratory toxicity
- allows more toxic Al species to bind to gill
- increases potential for ionoregulatory toxicity

E. Increasing Calcium in Water

- reduces Na^+ and Cl^- loss
- reduces Al binding to gill surface
- reduces ionoregulatory toxicity
- reduces gill damage
- increases diffusion distance
- increases potential for respiratory toxicity

Figure 2-3.—Diagram of fish gill and its relation to calcium (hypothesized relationships proposed by Gordon McDonald and Chris Wood, McMaster University, personal communication).

CAN LIMING HELP RESTORE OR PROTECT THE FISHERY?

To establish whether surface water acidity is the problem and whether liming is an appropriate management technique to enhance specific fisheries, the existing chemical, biological, and physical habitat conditions in the lake or stream should be evaluated to determine the specific factors that most limit the water's fisheries. In this chapter, we introduce the criteria to evaluate whether acidic water qualities limit fisheries. Next, we outline considerations for deciding whether fish population densities are less than expected for the type of water under review. Lastly, we briefly review physical habitat attributes that can limit fish densities in lakes and streams. Later in this manual, Chapter 6 introduces techniques useful for improving habitat conditions, while Chapter 7 provides a selection of methods for evaluating chemical, biological, and physical attributes in surface waters.

Figure 2-4 summarizes the criteria to consider when evaluating whether to lime a lake or stream. Responses to the first four criteria can result in clear decisions not to lime. We discuss reasons for these decisions more fully in this chapter. The next five criteria can lead to what we term "conditional liming." This means that the water could be limed with the knowledge that the treatment application will probably result in a limited period of effectiveness, or that only limited success is likely in achieving fisheries management goals. Even under these conditions, liming can still be used to achieve special or restrictive management objectives, such as protecting fish species of unique concern.

Liming is most effective and successful for fisheries management when conditions meet all nine of the criteria listed. For most purposes, two liming strategies are available— "mitigative liming" and "maintenance liming." Mitigative liming primarily treats surface waters having obvious and, often, recent impacts attributable to acidification. In contrast, maintenance liming generally is directed at preventing impacts where probable future effects are projected but have not yet occurred. Chapter 3 discusses these alternative liming strategies

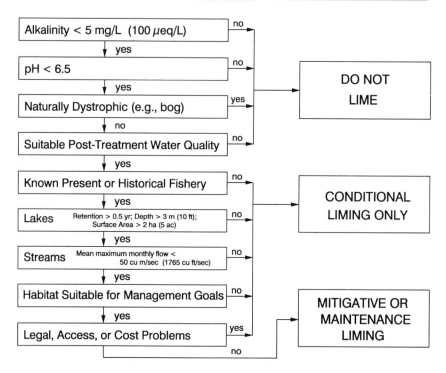

Figure 2-4.—Guideline criteria for determining needs and feasibilities of surface water liming.

and how to use them to achieve water quality and fishery goals established for liming programs.

Evaluating Water Quality Limitations

Alkalinity and pH

When evaluating whether to lime a lake, stream, or reservoir, historical surface water pH and alkalinity should be evaluated first (Fig. 2-4). Data collected should show that pH was less than 6.5 and that the alkalinity was less than 5.0 mg/L as $CaCO_3$ (100 µeq/L) either (1) frequently, (2) for two or more weeks per year, or (3) over seasonal times critical to sensitive life stages for important biota. Assuming that all alkalinity is associated with calcium, an alkalinity of greater than 5.0 mg/L indicates that the dissolved calcium concentrations should be greater than 2.0 mg/L, a concentration that would help limit toxic and low calcium physiological stress in most fish species.

Liming can help a fishery only when adverse water quality conditions affecting that fishery can be significantly altered by chemical treatment. Low alkalinity and high acidity are two easily measured water quality parameters that can distinguish potential problems associated with surface water acidification. As noted earlier in this chapter, alkalinity is a measure of the buffering capacity and an indicator of the acid neutralizing capacity present in surface waters. Alkalinities in the majority of lakes and streams in North America naturally range from 20 to 200 mg/L (400 to 4,000 µeq/L). Surface waters that are sensitive to potential acidification from atmospheric depositions generally have alkalinity levels less than 10 mg/L (200 µeq/L). Lakes and streams having such low alkalinities occur in many regions, including the northeastern United States, eastern Canada, Florida, the western mountains of North America, and northern Europe, especially the Scandinavian countries (Charles, 1991).

In addition to providing important information on sensitivity to potential acidification, alkalinity measurements also are useful indicators of the relative biological productivity potentials of these waters. Because they reflect mineral weathering rates, higher alkalinity concentrations generally indicate higher plant productivity in aquatic systems and higher food availability for fish and other aquatic animals. In highly productive North American hardwater lakes, alkalinities can range up to 400 mg/L or greater (Wetzel, 1983).

Alkalinity also provides a surrogate measure for the total concentrations of major alkaline-earth (calcium family) plus alkali (sodium family) metals in water. For most waters where liming may be beneficial, this primarily means the concentration of dissolved calcium. As discussed previously, calcium, in particular, is essential for the health of aquatic organisms; additionally, it can help mitigate effects of hydrogen ions and various heavy metals that can reach potentially toxic concentrations in acidified surface waters.

As also discussed before, decreasing pH levels below 6.0 increasingly stresses many aquatic species as a result of high hydrogen ion concentrations. In addition, the solubilities of many toxic metals increase and their more toxic forms become increasingly prevalent at lower pH levels, increasing the potential that metal toxicity will adversely affect aquatic life. Liming can mitigate many potentially adverse influences of surface-water acidification: it can increase pH, decreasing potentials for hydrogen ion toxicity; it can increase calcium concentrations, decreasing ionic stress and the permeabilities of many toxic metals to organisms; and it can decrease dissolved concentrations of many potentially toxic metals.

Naturally Dystrophic Surface Waters

Many naturally acidic (dystrophic) surface waters exist, including bogs and streams draining from bogs. Balanced biological communities often have evolved in these systems over a considerable time. Bog lakes often lack significant inlets or outlets, can have greater than 25 percent of the lake basin covered with floating mats of *Sphagnum* moss, contain brown water with an apparent color of greater than 75 platinum-cobalt units, and have dissolved organic carbon (DOC) concentrations of greater than 4.5 mg/L. Similar characteristics also can be associated with streams draining dystrophic systems. We recommend against liming such waters to prevent harming these specialized aquatic communities and upsetting the biological balance they have achieved (Fig. 2-4).

Lakes and streams where surface water acidification does not result from natural dystrophic processes but where acidification has persisted for an extended time also can develop communities of acidophilic ("acid-loving") biota. Since liming will produce chemical conditions less favorable to these biota, many such organisms will be lost from these systems following liming. For example, growths of *Sphagnum* would often disappear from shorelines of acidic lakes after liming. Decisions to lime must be made with the understanding that, after treatment, the biotic communities will revert to communities similar to, but not necessarily the same as, those existing before acidification.

Other Water Quality Limitations

Some acidic lakes also have water quality problems other than those primarily associated with acidity. Point and non-point sources of pollution can introduce toxicants into these waters, such as pesticides or other organic contaminants, that cannot be mitigated by liming. In addition, acid mine drainages or smelter emissions can continually release or deposit very high concentrations of heavy metals into surface waters and cause continual renewal of potentially toxic conditions (Penn Environmental Consultants, 1983). While liming strategies discussed in Chapter 5 can help mitigate impacts from some such sources, the procedures needed for successful mitigation can require frequent reliming or even continual liming to maintain suitable water qualities. Under the most extreme conditions of acid mine drainages, it may be impossible to maintain suitable water qualities by liming alone. Waters affected by point or nonpoint pollution that are not treatable by liming should be avoided (Fig. 2-4). Alternate water quality management tools may provide more successful fisheries management solutions in these situations (see, for example, Hammer, 1989).

In waters receiving certain chemical discharges (e.g., ammonia or waste organics) or lakes with excessive accumulations of organic matter, internally high chemical or biological demands for oxygen can develop. Such conditions can adversely affect fisheries by promoting summer or winter kills of fish. These waters, too, should be avoided when selecting appropriate lakes to lime, unless treatment also incorporates management to address the low oxygen conditions.

Finally, those surface waters that are most sensitive to acidification also contain very low concentrations of most dissolved ions, including phosphorus, nitrogen, and other nutrients required for plant growth. As noted previously, the low alkalinities found in surface waters sensitive to acidification also generally indicate low productivity potentials for fish and other aquatic life. Although liming can reduce stresses for aquatic life caused by acidification, it does not necessarily sustain highly productive fisheries. After removing limitations caused by acidic water chemistries, new limits can appear because of shortages of phosphorus, nitrogen, or another nutrient. Without additional water quality modifications to compensate for these new growth limitations, most fisheries in limed surface waters will not develop into highly productive or trophy-class fisheries.

Evaluating Fish Population Densities

Some presently acidic surface waters may not have supported fish populations even before they were acidified. Whether and to what extent liming can improve fisheries in these cases can be estimated from extant information about fisheries in these waters before their acidification. Consequently, a frequent requirement under many liming programs is that a surface water should have either an existing fish population (a self-sustaining, put-and-take, or put-grow-and-take fishery) or historical records showing that such a population previously existed in the water (Fig. 2-4).

Historic records for the resident fishery provide evidence that suspected acidification effects are actually occurring. As noted in this chapter, while recruitment failures can be an early effect for fisheries in acidifying waters, acidification-related causes are not the only reason for this phenomenon. It is preferable, therefore, to evaluate information on missing year classes in the light of what occurs in other surface waters within the region.

Sometimes, anglers may not have an accurate impression of the decline of fishing success in a lake or stream. Because acid-sensitive systems also tend to have naturally low productivities, low catch rates may just reflect low productivity rather than any acid-related effects. If such relationships are unclear, more intensive investigations of the angling success and standing crops of fish may be warranted.

The history of many fisheries commonly is recorded though regular or occasional "creel surveys" of angler catches. Comparative information from past and recent creel surveys can be used to reveal trends in overall angling effort and success as well as fundamental changes in the population structure, as shown through changes in the composition of the angler catch.

To be valid, the results compared must come from surveys using similar sample times and durations and similar computation techniques.

Chapter 7 provides additional information on the use and implementation of creel surveys. It also introduces other methods that can provide direct information on the present status of fisheries in lakes, streams, or reservoirs.

Evaluating the Physical Habitat

In addition to the chemical and biological considerations discussed previously, physical habitat characteristics can substantially limit the potential effectiveness of liming and place additional limitations on fisheries in treated waters. Sometimes, these habitat constraints may preclude reaching some management goals for fisheries.

Three physical conditions should be met to achieve the greatest potentials for successful limings of lakes and reservoirs (Fig. 2-4). The first concerns the rate of water flow. Because the flow of water through lake basins flushes out the residual neutralizing effects of limestone along with the treated waters, the hydraulic retention time for a lake or reservoir should be greater than six months. Liming frequencies greater than once per year may be necessary if average water retention time is less. Conversely, longer retention times lengthen the period between necessary retreatments. For example, a lake with a water retention time of five years may not require reliming for six years or longer (see Fig. 5-5).

Average retention rates for some lakes have been calculated by the U.S. Geological Survey (USGS) and by various state and local agencies. When this information is not available, these rates must be calculated. Average hydraulic retention time is the inverse of flushing rate (1/flushing rate). To calculate the average flushing rate for a lake (i.e., the number of times per year the water is exchanged through a lake), multiply the watershed area (hectares or acres) by the mean annual runoff (meters or feet) per year, then divide by the total lake volume (hectare–meters or acre–feet). Chapter 7 discusses methods to calculate lake volume.

Another method to calculate flushing rates is to divide the average annual discharge rate from the outlet stream(s) of the lake into the lake volume (using the same volume units for both measurements). Because of wide variations in evaporation and soil percolation rates, no predictable relationship exists between annual precipitation and runoff rates.

The second physical characteristic concerns water depth. Lakes with depth less than 3 m (10 ft) or surface-water areas less than 2 ha (5 ac) generally have limited habitat availabilities and low fishery potentials. Such lakes can often develop seasonally low oxygen concentrations and experience winter or summer kills. As such, lakes not meeting these minimum depth and area requirements should be limed only to meet special objectives, with the recognition that the potential duration and effectiveness of the treatment will be limited. Lakes where this strategy can be appropriate include those that are managed for high yield, put-and-take recreational fisheries; for example, lakes located in or near metropolitan areas.

The water flow rate also affects decisions on whether to lime a stream. When liming streams, the maximum mean monthly flow should be less than 50 m^3/sec (1,765 ft^3/sec). Greater flow rates make it extremely difficult to deliver the mass of limestone required for adequate neutralization. To attain the greatest potential benefits, the stream reach should extend at least 3 km (2 miles) downstream of the treatment point.

The third physical characteristic to determine when considering liming is whether the physical habitat in the treated system can adequately support the fishery goals set by the management plan. For example, if the management goal calls for establishing a self-sustaining fishery, does adequate spawning habitat exist? Chapter 3 introduces ways to define reasonable water quality and fishery management goals for treated waters. Chapter 6 discusses methods for assessing potential fishery and habitat limitations in the treated lake or stream and suggests alternatives that can be used to lessen these limitations.

IS LIMING A FEASIBLE MANAGEMENT OPTION?

After completing the suggested evaluations, the ultimate decision whether to lime will depend on the cost and the resources available. Although treatment costs vary widely, depending on whether the water to be treated is a pond, lake, or stream, some general estimating guidelines follow for making an initial cost evaluation. Chapter 5 contains detailed cost information for limestone materials and additional sources of information.

Costs Associated with Liming

Estimating the costs of liming will depend on the actions required to test, treat, monitor, and, perhaps, re-treat the body of water. After determining the costs of liming, you must consider a variety of other factors, including initial delivery and application of the lime; monitoring the water chemistry and sampling the biological factors after liming; and reapplication when necessary.

Estimating the initial application cost factors will depend in part on who conducts the activities. Resource management agencies can generally use in-house labor. However, lake owner associations or similar private agencies may use volunteer labor for the application but find it necessary to hire consultants to conduct chemical and biological monitoring of the results. And unless the source of acidity to the waterbody is controlled and its reacidification abated, the cost of repeating the process must also be included when estimating the costs of liming.

One constant, however, will be the cost for and delivery of the liming materials to the liming site. For the application of

Table 2-1.—Typical costs for application of limestone materials to lakes, 1991.

APPLICATION METHOD	COST PER SURFACE HECTARE	COST PER ACRE
Helicopter	$200–650	$80–260
Barge with delivery system	50–350	20–140
Slurry box in boat	50–350	20–140

limestone materials to lakes, the costs listed in Table 2-1 are generally applicable.

The cost of maintaining water quality in a limed lake is determined principally by how fast the lake flushes. If a lake completely flushes its volume in one year, treatment will be effective for at least two and, at most, four years, depending on annual rainfall. For a given lake volume, the cost per flush is essentially constant, provided certain treatment conditions are met. Limestone materials should be high in calcium, of proper size, and, usually, slurried prior to use. The United States has a great many sources of suitable, high-calcium limestone (discussed in Chapter 5), making it readily available. The cost of transporting limestone varies little from state to state. However, depending on the size of the lake or quantity to be used, the costs will vary.

The Living Lakes, Inc., program listed its experience in liming 36 lakes ranging in size from 4 to 140 ha (10 to 350 ac). Indications are that those lakes that flush slower than once every eight months can be managed effectively for acidity by direct lake application. The pH in these lakes varied from 4.5 to 6.0. Annual costs for a broad range of lake sizes, including minimal monitoring, fell within a fairly narrow range. For lakes in the LLI program, annual treatment costs ranged from $400 to $3,000, depending upon the type of application (Table 2-2).

For lakes that flush more rapidly than eight months, or where fish spawn in stream tributaries, liming the soil of the whole or partial watersheds may compete economically with direct lake application. Based upon Scandinavian and British soil liming experiences, costs range from $160 to $320/ha/yr ($400 to $800/ac/yr). Estimates are that water quality may be improved over a 15- to 20-year period. If the results of the few soil limings prove to be transferable to most watersheds, such treatment appears to last longer than direct lake liming and protects lake tributary streams and forest soils in an effective, low-technology manner.

The cost for stream liming varies and depends upon the characteristics of the stream, including its water quality and the quantity to be treated. One example is a 11 km (7-mile)

Table 2-2.—Comparative annual costs for liming lakes of different sizes.[a]

LAKE DATA			ANNUAL COSTS				SUMMARY
TYPE OF APPLICATION	SIZE[b] (ACRES)	EFFECTIVE-NESS (IN YEARS)	TREATMENT	MONITORING	OVERSIGHT	TOTAL	COST/ACRE/YR[c]
Slurry box	43	3.5	$766	$80	$225	$1,071	$25
Slurry box	10	3.5	430	80	145	655	66
Slurry box	130	6.1	579	40	90	709	5–6
Slurry box	21	1.8	1,705	220	600	2,525	120
Barge	322	5.8	2,193	80	450	2,723	8–9
Barge	133	6.1	731	55	125	911	7
Barge	212	7.2	813	40	100	953	4–5
Barge	49	5.5	1,634	110	375	2,119	43

[a] Modified from Living Lakes, Inc. 1989 Annual Report, Washington, DC.
[b] To convert to hectares, multiply by 0.404.
[c] To convert to cost/hectare/yr, multiply by 2.47.

stretch of a small stream with an average flow of about 0.014 m^3/sec (0.5 ft^3/sec) and a width of approximately 3–6 m (10–20 ft). The management objective was to improve water quality so that brook trout could reproduce and restock a former fishery. Volunteers constructed a delivery unit costing approximately $20,000, and annual material costs are about $9,000. Qualified volunteer professionals monitor the effects. These cost estimates are broad and vague, but they can give the reader a general feeling for relative costs.

CHAPTER 3

Setting Reasonable Fishery and Water Quality Management Goals

When liming is selected as a feasible management technique for improving water quality conditions in a lake, stream, or reservoir, the next step is establishing appropriate fishery and water quality management goals for that waterbody. Most surface waters are dynamic systems with continual renewal of water and continual loss of treated waters. Therefore, once a surface water is limed as part of a fishery management program, a continuing need for monitoring and periodic future retreatment remains. Because liming is a long-term management commitment, the management team must establish reasonable, achievable goals. Setting reasonable fishery and water quality goals for treated surface water requires specific consideration of the various limitations present.

Most liming programs achieve their fishery management goals using one of two primary treatment plans:

1. **Mitigative liming** — restoring a fishery that has been damaged or completely lost as a result of acidification; or

---

2. Maintenance liming — maintaining or protecting a fishery threatened but not yet damaged by acidification.

Table 3-1 shows general guidelines that can be used to establish appropriate objectives using each type of treatment plan. When environmental conditions fall outside those presented in Table 3-1, liming will be less effective in the long term. Nevertheless, liming still can achieve short-term or intermediate benefits for some lakes and streams where other options are lacking. Liming under these conditions is particularly useful to reduce threats from acidic waters or elevated metal concentrations to fishery resources of particularly unique or economic importance. This chapter discusses the guidelines in Table 3-1, introducing the considerations necessary to establish reasonable goals for water quality and fishery management of treated waters.

WATER QUALITY GOALS

The primary goal of the surface water liming part of an overall fishery management program is to improve or maintain water quality that achieves fisheries management objectives for the longest possible time. Based on the information in Chapter 2, this can mean increasing the pH of the water to greater than 6.5 and the alkalinity to greater than 5 mg/L as $CaCO_3$ (100 µeq/L).

Figure 3-1 shows the critical pH range for 25 different fish species for which appropriate data are available. The information in this figure was compiled by screening all appropriate and available laboratory and field information, including presence and absence data for lakes (Baker and Christensen, 1991). Since the relationships shown include data from field observations of the effects on population levels, responses by all life stages are implicit. Fishery management can apply these relationships to define the minimum pH level at which potential effects from acidic water threaten each fish species. Figure 3-1 provides general guidelines, by fish species, for the minimum pH values at which retreatment is necessary (e.g., 4.8 for

Table 3-1.—Site selection guidelines for mitigative and maintenance liming of lakes, streams, and reservoirs.

LIMING GOAL	SELECTION GUIDELINES
Mitigative Liming	
Biological Status	The water must have an existing fish population (naturally reproducing, put-and-take, or put-grow-and-take), or it must have historical records of fish presence. If fish are present, deleterious effects should be apparent. Any such effects should not be the result of nutrient enrichment, point source pollution, interspecific competition, or fishing pressure.
pH/Alkalinity	Surface water pH < 6.0 or alkalinity < 0.5 mg/L (10 μeq/L) should have occurred in the water sometime during the past 5 years either (1) frequently, (2) for 2 or more weeks, or (3) during critical periods for sensitive life stages of important aquatic species.
Critical Habitat	Suitable habitat must exist for the species of interest. If stress to acid sensitive species has been documented, this stress should not have been caused by damage or loss of suitable habitat.
Pollution Sources	The water should not be affected by any point or non-point pollution sources that would interfere with maintaining healthy biological communities following liming.
Hydrology	Lakes and reservoirs should have a retention time of at least 0.5 year, unless stream or watershed treatments are used. Streams should have maximum monthly flows of < 50 m^3/sec (1,765 ft^3/sec).
Size	Lakes and reservoirs should have a surface area > 2 ha (5 ac) and a maximum depth of > 3 m (10 ft). Streams should extend at least 3 km (2 miles) downstream to maximize potential benefits from treatment.
Maintenance Liming	
Biological Status	The water must presently support populations or communities of fish having special ecological or recreational importance. If these populations or communities show signs of stress, these effects should not be caused by nutrient enrichment, point source pollution, interspecies competition, or excess fish pressure.
pH/Alkalinity	Surface water pH < 6.5 and alkalinity < 5 mg/L (100 μeq/L) should have occurred in the water sometime during the past 5 years either (1) frequently, (2) for 2 or more weeks, or (3) during critical periods for sensitive life stages of important aquatic species.
Critical Habitat	Same as mitigative liming.
Pollution Sources	Same as mitigative liming.
Hydrology	Same as mitigative liming.
Size	Same as mitigative liming.

27

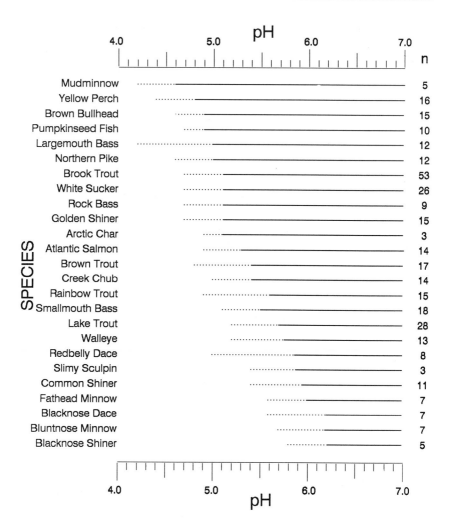

Figure 3-1.—Estimated critical pH values for adult fish populations based on available literature for field and laboratory studies (dashed lines indicate approximate range of uncertainty for estimated critical pH; modified from Baker and Christensen, 1991).

yellow perch). It also gives guidelines for selecting fish species with low sensitivities to acidic water qualities for fishery management in waters where liming is not practical.

A brief example can demonstrate how the information in Figure 3-1 may be used. Anglers at Big Clam Lake complained that brook trout fishing is not as good as it used to be. "The lake is acid," they say, "and we don't see any of the minnows that used to be abundant along the shore." Examination of the creel survey records shows that angling pressure has declined slightly in the past 4 years, compared to the previous 10 years when total angling pressures remained relatively constant. These recent records also show that catch per unit effort by anglers has decreased. After evaluating this information, the local fisheries biologist completed field chemistry analyses and determined that the pH was 5.0 and alkalinity equaled zero. Referring to Figure 3-1, we see that in acidic waters with pH levels less than 5.2, brook trout have an uncertain existence. Based on water quality considerations alone, liming could improve water quality conditions for brook trout in this lake. The target pH for liming to achieve was set at 6.5. Moreover, when the lake water pH again begins to decline below about 5.5 following liming, the lake should be relimed.

FISHERY GOALS

Fishery management goals in limed surface waters will usually include maintaining a fishery as similar as possible to the historical fishery populations and communities resident in the water. This frequently will include establishing or maintaining fish populations with natural reproduction rates adequate to sustain harvests that satisfy angler demands. Available management strategies can lead to establishing put-and-take fisheries, put-grow-and-take fisheries, or self-sustaining fisheries. Table 3-2 lists a few representative publications that contain useful considerations for managing fish in small lakes and ponds. Similar publications are available from agriculture extension offices in most states.

Establishing appropriate management goals and strategies for fisheries within management programs that use liming should include considerations in addition to water quality changes. These considerations include assessing the current status of the fishery, selection of appropriate fish species, and ecological limitations posed by the environment and stocking.

Table 3-2.—Representative publications for the culture and management of fish populations in small lakes and ponds.

NEW YORK

Epper, A.W., H.A. Regier, and D.M. Green. 1988. Fish Management in New York Ponds. Inf. Bull. 116, Coop. Ext. Serv., Cornell Univ., Ithaca.

PENNSYLVANIA

Pennsylvania Cooperative Extension Service. 1984. Pennsylvania Fish Ponds. College of Agric., Penn. State Univ., University Park.

MICHIGAN

Schrouder, J.D., C.M. Smith, P.J. Rusz, and R.J. White. 1982. Managing Michigan Ponds for Sport Fishing. Ext. Bull. E1554, Coop. Ext. Serv., Michigan State Univ., East Lansing.

WISCONSIN

Klingbiel, J.H., L.C. Sticker, and O.J. Rongstad. (undated). Wisconsin Farm Fish Ponds. Manual 2, Coop. Ext. Prog., Univ. Wisconsin, Madison.

KANSAS

Gabelhouse, D.W. Jr., R.L. Hager, and H.E. Klaassen. 1987. Producing Fish and Wildlife from Kansas Ponds. 2nd ed. Kansas Dep. Wildl. Parks, Pratt.

COLORADO

Satterfield, J.R. Jr. and S.A. Flickinger. 1984. Colorado Warmwater Pond Handbook. Fish. Bull. No. 1, Colorado State Univ., Fort Collins.

CALIFORNIA

Calhoun, A., ed. 1966. Inland Fisheries Management. Dep. Fish and Game, The Resour. Agency, State of Calif., Sacramento.

CHARACTERIZING PRESENT FISHERY AND HABITAT CONDITIONS

The status of the fishery prior to liming should be clearly characterized. This information can help determine if fisheries are affected by acidic conditions and establish baseline conditions to measure the benefits of liming. For example, as discussed in the previous chapter, fisheries affected by increasing acidification often show increased frequencies of recruitment failures, revealed by the absence of whole year-classes of fish in the population structure. When effects are prolonged, recruitment failures can lead to a preponderance of older fish and, with extreme impact, eradication of the entire fishery. Such patterns, coupled with records of pH decreasing to levels at the lower population tolerance limits for the species (Fig. 3-1), provide strong evidence that acidity is affecting the fish population.

If patterns for fish populations or pH are different, then alternate causes of poor fishing in the lake or stream should be evaluated. For example, has the water level of the lake changed in recent years, or has access to spawning areas or a tributary stream been blocked? Many lakes and streams lack adequate spawning or rearing habitat, preventing the establishment of a self-sustaining population for many fish species. Assessment of existing habitat conditions prior to liming can reveal whether modifications to the lake or stream, in addition to or instead of liming, may greatly benefit fisheries. Chapter 7 presents specific methods for assessing those aspects of fisheries in lakes and streams, and for assessing and improving various habitat characteristics for fisheries.

SELECTING TARGET FISH SPECIES AND ECOLOGICAL LIMITATIONS

Under many liming programs, the fish species targeted in the management goals include the same fish species currently inhabiting the lake. When no existing fishery is present, the post-liming management goals for fisheries can be based on those populations that historically inhabited the water. Similarly, post-liming management goals on whether to manage for put-and-take, put-grow-and-take, or self-sustaining fisheries can be based on which kind of fishery historically resided in the water. Such information is often available from local offices of state fisheries agencies. For some waters, however, the only source of historical fisheries information may be old newspaper clippings, lake association newsletters and memoranda, or just anecdotal accounts from older anglers in the area.

If information on the residence of historical species is lacking or introduction of a new fishery is part of the management program, selecting species requires special consideration. When introducing new species, we recommend using those that have relatively low sensitivity to acidic conditions. Fish that are less sensitive to acid water qualities will maintain healthier populations when the water begins to reacidify. The information in Figure 3-1 can help identify such species.

Designing a fishery management plan requires specific consideration to selecting species not excessively limited by ecological conditions present in the treated habitat. To evaluate potential limitation of habitat conditions for candidate fish species (see procedures in Chapter 7), consider any special requirements for the separate life stages of each fish species considered in addition to the pH information in Figure 3-1.

Table 3-3 presents some specific habitat requirements that are important limitations for several fish species. These requirements include spawning, temperature, pH, and calcium needs. This table shows considerations for determining and managing water quality and habitat requirements for successful fisheries of the most commonly encountered game fish. Additional information on life history requirements and potential habitat limitations is available in a series of reports produced by the U.S. Fish and Wildlife Service that summarize habitat suitability requirements for over 30 fish species (Table 3-4). Region-specific information on habitat requirements and unique management considerations can be obtained from local offices of state fishery management agencies.

The listed fish species can be divided roughly into cold- and warmwater species. Coldwater species, which include Atlantic salmon, brown trout, brook trout, lake trout, and rainbow trout, can tolerate temperature ranges of below -1°C to 27°C (30°F to 80°F) but generally prefer temperatures below 18°C (65°F). These fish are routinely found in lakes and streams in their northern range and in deeper lakes in the southern portions of their ranges. They feed primarily on aquatic and, occasionally, terrestrial insects as young, and on forage fish and smaller game fish as adults. Depending on geographic location, water temperature, and fish species, they spawn from early fall through late spring. All species except lake trout spawn over gravel or rocky streambeds or lake inlets or outlets. Lake trout spawn over lake bottoms covered with rock and/or rubble. Some researchers believe that brook trout, which can spawn on either stream or lake bottoms, require gravel and spring-fed or other upwelling waters.

Table 3-3.—Life stage requirements for specific fish species.

COMMON NAME	SCIENTIFIC NAME	SPAWNING REQUIREMENTS	WATER TEMPERATURE REQUIREMENTS	pH REQUIREMENTS (EGGS, JUV., ADULT)	CALCIUM REQUIREMENTS
Atlantic salmon, landlocked	*Salmo salar*	Gravel riffles of inlet and outlet streams	Below 21°C (70°F)	Above 6.0	Above 2.5 mg/L
Brown trout	*Salmo trutta*	Stream gravel ranging in size from 1/4 to 3 inches in diameter with good vertical water flow	Tolerant from 0°C–27°C (32°F–80°F)	Above 6.0	Above 2.5 mg/L
Lake trout	*Salvelinus namaycush*	Lake bottoms over rock and rubble	Prefer temps. below 13°C (55°F)	Above 6.0	Above 2.5 mg/L
Brook trout	*Salvelinus fontinalis*	Spring-fed areas on lake bottoms; gravel beaches with up-welling seep water; gravel riffles on lower ends of pools in streams	Tolerant from 0°C–24°C (32°F–75°F)	Above 6.0–9.5	Above 2.5 mg/L
Rainbow trout	*Oncorhynchus mykiss*	Stream riffles and lower ends of pools; may spawn in lake outlets	Tolerant from 0°C–27°C (32°F–80°F)	Above 6.0; tolerant from 5.8–9.5	Above 2.5 mg/L
Striped bass, landlocked	*Morone saxatilis*	Upper tidewater reaches of freshwater rivers that have considerable current	Tolerant from -1°C–32.5°C (30°F–90°F)	Above 6.0	Above 2.5 mg/L
Largemouth bass	*Micropterus salmoides*	Substrate of sand, gravel, roots, or aquatic vegetation	Above 27°C (80°F)	Above 6.0	Above 2.5 mg/L
Smallmouth bass	*Micropterus dolomieu*	Substrate of sand, gravel, or rocks	21°C–27°C (70°F–80°F)	Above 6.0	Above 2.5 mg/L
Walleye	*Stizostedion vitreum*	Eggs broadcasted in 12- to 30-inch deep water over gravel riffles or gravel shoals	Tolerant from 0°C–32.5°C (32°F–90°F)	Above 6.0	Above 2.5 mg/L
Yellow perch	*Perca flavescens*	Eggs spawned in a gelatinous matrix near shore; egg strings woven around aquatic plants or brush	Tolerant from 0°C–32.5°C (32°F–90°F)	Above 6.0	Above 2.5 mg/L

Table 3-4.—Habitat suitability index models for freshwater fishes.

COMMON NAME	SCIENTIFIC NAME	NTIS NUMBER[a]	USFWS NUMBER[a]
Acipenseridae			
Shortnose sturgeon	*Acipenser brevirostrum*		FWS/OBS-82/10.129
Polyodontidae			
Paddlefish	*Polyodon spathula*		FWS/OBS-82/10.80
Salmonidae			
Arctic grayling	*Thymallus arcticus*		FWS/OBS-82/10.110
Brook trout	*Salvelinus fontinalis*	PB83-147041	FWS/OBS-82/10.24
Brown trout	*Salmo trutta*		FWS/OBS-82/10.124
Chinook salmon	*Oncorhynchus tshawytscha*		FWS/OBS-82/10.112
Coho salmon	*Oncorhynchus kisutch*	PB84-131150	FWS/OBS-82/10.49
Cutthroat trout	*Oncorhynchus clarki*	PB82-239922	FWS/OBS-82/10.5
Lake trout	*Salvelinus namaycush*		FWS/OBS-82/10.84
Rainbow trout	*Oncorhynchus mykiss*		FWS/OBS-82/10.60
Esocidae			
Northern pike	*Esox lucius*	PB83-143164	FWS/OBS-82/10.17
Cyprinidae			
Blacknose dace	*Rhinichthys atratulus*	PB84-129501	FWS/OBS-82/10.41
Common carp	*Cyprinus carpio*	PB83-150615	FWS/OBS-82/10.12
Common shiner	*Notropis cornutus*	PB84-128529	FWS/OBS-82/10.40
Creek chub	*Semotilus atromaculatus*	PB82-239914	FWS/OBS-82/10.4
Longnose dace	*Rhinichthys cataractae*	PB84-118090	FWS/OBS-82/10.33
Catostomidae			
Bigmouth buffalo	*Ictiobus cyprinellus*	PB84-129469	FWS/OBS-82/10.34
Longnose sucker	*Catostomus catostomus*		FWS/OBS-82/10.35
Smallmouth buffalo	*Ictiobus bubalus*	PB83-143172	FWS/OBS-82/10.13
Ictaluridae			
Channel catfish	*Ictalurus punctatus*	PB82-229519	FWS/OBS-82/10.2
Black bullhead	*Ictalurus melas*	PB83-147025	FWS/OBS-82/10.14
Percichthyidae			
Striped bass	*Morone saxatilis*	PB82-237447	FWS/OBS-82/10.1
Centrarchidae			
Black crappie	*Pomoxis nigromaculatus*	PB82-239930	FWS/OBS-82/10.6
Bluegill	*Lepomis macrochirus*	PB82-239955	FWS/OBS-82/10.8
Green sunfish	*Lepomis cyanellus*	PB83-148205	FWS/OBS-82/10.15
Largemouth bass	*Micropterus salmoides*	PB83-142497	FWS/OBS-82/10.16
Smallmouth bass	*Micropterus dolomieu*	PB84-129451	FWS/OBS-82/10.36
Warmouth	*Lepomis gulosus*		FWS/OBS-82/10.67
White crappie	*Pomoxis annularis*	PB82-239948	FWS/OBS-82/10.7
Percidae			
Slough darter	*Etheostoma gracile*	PB82-242322	FWS/OBS-82/10.9
Walleye	*Stizostedion vitreum*		FWS/OBS-82/10.56
Yellow perch	*Perca flavescens*		FWS/OBS-82/10.55

[a] Available from the National Technical Information Service (NTIS), U.S. Department of Commerce, 5285 Port Royal Rd., Springfield, VA 22161.

Striped bass, largemouth and smallmouth bass, northern pike, walleye, and yellow perch inhabit warmer waters than trout and salmon. In this group, landlocked striped bass are unusually adaptable to subtropic U.S. waters. While all these species occur in lakes and streams, largemouth bass are more often found in standing water habitats. Water temperatures tolerated by these fish species range from 0°C to 32.5°C (32°F to 90°F), although the preferred temperature range for these species is generally 18°C to 27°C (65°F to 80°F). The young of these species feed primarily on crustaceans and invertebrates, while adults feed mainly on insects, fish, and some semi-aquatic vertebrates (e.g., frogs). These species spawn from spring through early summer. Largemouth and smallmouth bass, walleye, and striped bass spawn over substrates of sand, gravel, rocks, and, occasionally, aquatic vegetation. Yellow perch are unique among freshwater fish; they lay eggs near shore in gelatinous, accordion-like strings, which can swell to eight feet in length.

As stated in Chapter 2, most waters sensitive to the potential effects of acidification have very low natural biological growth rates. Because of these inherent limits on productivity, it is likely that very few limed waters can be managed to generate "trophy fisheries." These naturally low productivities also place additional limits on selecting appropriate species combinations for limed waters because low productivities also limit the complexity of food chains possible in these aquatic systems. Therefore, the most successful fisheries in limed waters will often consist of one or, at most, two or three fish species. If combinations of two or more fish species are included in the management plan, the second and third or, rarely, fourth species are usually selected as forage species to provide prey for the primary or target species.

Selecting these additional species requires extreme caution to assure they will not compete for food or habitat space with the target fish species. For example, many forage species prey on the insects and other invertebrate food sources used by various life stages of the target trout species. Sometimes, larger forage species may even prey on the smaller life stages of the target species. In Table 3-5, we suggest some acceptable

Table 3-5.—Appropriate species combinations for managed surface waters.

WATER	PREDATOR	PREY	CONCERNS
Warm	Largemouth bass	Bluegill	Not recommended for northern states because bluegill overpopulate.
Warm	Largemouth bass	Fathead minnow	
Warm	Largemouth bass	Black crappie	Successful only if high density of large bass is maintained.
Warm	Smallmouth bass	Fathead minnow	
Warm	Channel catfish	Fathead minnow	
Cold	Rainbow trout	Invertebrates	Forage fish compete with small trout for food.
Cold	Brook trout	Invertebrates	Forage fish compete with small trout for food.
Cold	Rainbow trout	Invertebrates	Forage fish compete with small trout for food.
Cold	Lake trout	Invertebrates & other fish	Forage fish compete with small trout for food.

combinations of fish species. Specific information concerning other appropriate fish species combinations of regional importance is also available from local offices of state fishery management agencies.

ADDITIONAL FISHERY MANAGEMENT CONSIDERATIONS

When selecting target fish species and setting goals under fishery management and liming programs, angler demands must be considered. That is, should or can anglers get what they want? We previously mentioned that most limed waters provide very limited potential for establishing trophy fisheries. Thus, anglers should be discouraged from expecting such possibilities. On the other hand, although yellow perch and brown bullheads have relatively low sensitivities to acidic water qualities, they also have limited appeal to many anglers. So these species may not be the best choice for some waters. Target species must have local importance to anglers or the public in general.

Natural limits on productivities in limed surface waters also limit growth potentials under put-grow-and-take and self-sustaining fishery management strategies. Consequently, when using these options, expectations about growth performances should be lowered and management actions can include establishing special angling and harvest restrictions. Chapter 6 presents some useful management approaches to achieve these goals.

Finally, when habitat constraints limit the maintenance of self-sustaining fish populations, augmenting or even maintaining the designated species entirely by stocking may be necessary. Also, where mitigative liming is conducted with plans to restore historical fisheries that have been substantially reduced or lost because of acidification, stocking can be essential. Appropriate considerations on selecting fish stocks, size classes, stocking times, and stocking densities are introduced in Chapter 6.

CHAPTER 4

Obtaining Legal Permits

C hemical alteration of lakes and streams is a serious undertaking that is highly regulated by municipal, state, and federal agencies charged with protecting the environment. Regulations vary with the proposed activity and may differ depending upon whether the receiving water is public or private. Permits may (or may not) be required to apply liming material in a private lake or stream. However, a permit will almost always be necessary to stock fish because existing fisheries must be protected from potential damage or decimation from exotic species through disease introduction, competition for food and spawning space, and predation.

The necessity for a permit to apply limestone materials to receiving waters also varies among states and on federal land within different states. When a decision is made to lime a body of water and perhaps stock that waterbody with fish, the prudent course of action is to seek a permit. Information on whether a permit is required can be quickly learned. If a permit is required and not obtained, serious penalties may be assessed. Although the issuing agencies and the permit process vary from state to state, the guidelines in this chapter will be helpful when planning a liming project.

Several states have established liming policies. For example, New York and Massachusetts have prepared environmental impact statements specifically dealing with the liming of surface waters (Mass. Div. Fish. Wildl. 1984; Simonin et al. 1990). As a result, both states have developed guidelines and criteria to permit liming. In New York, the permit process is handled by the Department of Natural Resources for both public and private waters outside of the Adirondack Park. Within that park, the Adirondack Park Agency has jurisdiction and issues permits for liming and the Department of Natural Resources, Fisheries Division, issues permits for stocking of fish.

In West Virginia, liming is handled on a case-by-case basis through the Department of Natural Resources, which may not require a permit of any type. Some investigation may be necessary to find the appropriate agency to either issue the permit or to state that one is unnecessary. In any case, certain types of information will be required to obtain a permit to lime and to stock fish. However, the process that evaluated chemical, physical, and biological aspects of the project to determine its viability will usually provide the necessary data.

In most cases, an operational treatment plan must be assembled and filed with the permit application. Sometimes the plan itself will serve as the application. Generic information often required for an application includes:

- **Baseline Water Quality**
  - *Temperature*
  - *pH*
  - *Water color and dissolved organic carbon*
  - *Conductivity*
  - *Alkalinity or acid neutralizing capacity*
  - *Major nutrients (phosphorus, nitrogen, calcium)*
  - *Metals (aluminum, mercury, etc.)*

- **Physical and Hydraulic Data**
  - *Surface area*
  - *Depth*
  - *Volume*
  - *Retention time (lake)*
  - *Flow rate (stream)*
- **Aquatic Vegetation**
  - *Type*
  - *Amount*
  - *Location (if not uniformly distributed)*
  - *Substrate composition (sometimes for streams)*
- **Treatment Material Documentation**
  - *Chemical composition (percent calcium, percent other)*
  - *Type of material (size—e.g., pellets, powder, slurry)*
  - *Amount to be applied*
  - *Method of application (boat, barge, helicopter, doser)*

The site-specific treatment plan must clearly outline the liming project's objective. That is, it must identify the site by map coordinates, state the current water quality and the water quality to be achieved by the liming, the time of year the liming treatment will take place, the time required to apply the liming material, the plan for documenting the changes in water quality as a result of the liming, and the plan for monitoring the water quality for potential re-liming. This plan should also include whether fish will be stocked; however, detailed information will be submitted to the appropriate agency to procure a permit. Additional aspects include general anecdotal information about previous experience with the planned type of treatment, whether the lake is surrounded with dwellings, whether it previously had populations of fish, and so forth. If the lake or stream is not remote, the plan should include information on how local residents and responsible officials will be informed of the liming and its associated activities.

In summary, the application and treatment plan should give the reviewer enough information to ensure that the integrity of the lake or stream will be protected and that the liming will not cause harmful effects, such as locating a doser in a wetland where it can affect sensitive biota. Figure 4-1 and Table 4-1 illustrate some of the information necessary and a way to present it.

Little Simon Pond

New York

Figure 4-1.—Map showing zones for application of limestone to Little Simon Pond. Numbers refer to depth of water in feet.

An outline of the fishery management plan must be submitted and arrangements completed before liming. Authorities can refuse a stocking permit even though the water is treated and its quality is conducive to fisheries enhancement. They may find, for example, that the goals of the fishery plan do not fit into overall state or federal goals for regional waterbodies.

Table 4-1.—Basic treatment information for Little Simon Pond.

LOCATION:	TUPPER LAKE, NY	SITE ID: NYL001
CONTACT:	J.E. DOE (owner)	M.R. SMITH (caretaker)
	518/555-1234	518/555-5678

TREATMENT DATES:
Primary: 6, 7, 8 August; 0600 mix batch; 0800–1900 apply slurry
Alternate: 7, 8, 9 August; 0600 mix batch; 0800–1900 apply slurry

STAGING AREA:
Primary: Tupper Lake Rod and Gun Club

DOSAGE:

Dry dose: (tons)	EcoCal	14	25 tons
	EcoCal	18	50 tons
	ATF	40	25 tons
	TOTAL		100 tons
Water required:	6,000 gal.		

DOSE BY ZONE (see figure 4-1)
DRY DOSE (TONS)

ZONE	ACRES	EcoCal 14	EcoCal 18	ATF 40	TOTAL
1	13.3	1.9	3.8	2.5	8.2
2	10.3	1.4	2.8	1.5	5.7
3	20.1	7.7	15.7	11.5	34.9
4	11.5	3.0	6.2	7.5	16.7
5	16.4	2.0	3.7	0.0	5.7
6	19.8	2.2	4.3	0.0	6.5
7	23.1	3.3	6.5	2.0	11.8
8	41.5	3.5	7.0	0.0	10.5
TOTAL	156.0	25.0	50.0	25.0	100.0

Consequently, coordinating liming and stocking plans with the responsible authorities before applying for a liming permit is vital.

After a successful liming, appropriate species of fish may be procured and stocked in the lake or stream. The best sources of information on fish hatcheries certified to supply acceptable species for stocking are the local offices of state and federal fisheries management agencies. Charged with regulating the production, growth, and distribution of fish in their states, these agencies know which hatcheries can supply high quality, certified fish. First, however, these same fisheries agencies will require information about the purposes and goals of the stocking and fisheries management plan. The following information is the minimum for obtaining a permit to stock fish:

- Location of waterbody (size, hydrology)
- History of fishery
- Type of fishery (warmwater, trout)
- Current condition of the fishery
- Species of fish to be stocked
- Number to be stocked
- Sources of fish

Most state fisheries agencies will help in the development of a reasonable management plan for the fisheries. However, the applicant should prepare a basic suggested plan for the agency to analyze because resources at most state agencies are limited and assigned to the management of public fisheries.

Extension agents at state universities can help, and environmental consulting groups that specialize in lake and stream remediation and fisheries management may offer services. Chapter 5 provides additional information on developing appropriate fish stocking programs, and Chapter 9 contains examples of site treatment plans for a lake, a stream, and a watershed.

CHAPTER 5

Obtaining and Applying Limestone

WHY USE LIMESTONE RATHER THAN OTHER MATERIALS?

Many alkaline materials can neutralize acidic surface waters, such as limestone, hydrated or calcined lime, soda ash, sodium bicarbonate, and sodium hydroxide. In recent years powdered limestone has been recognized as the best material for managing mildly acidic surface waters. Compared to other materials, limestone is easier to handle, more widely available, and less expensive, can provide a sediment dose, and is less likely to increase pH to levels that may harm aquatic life. Therefore, only the use of limestone is discussed in this chapter. The reader is referred to Fraser and Britt (1982) and Olem (1991) for other materials that may be more appropriate for treating highly acidic waters; that is, waters with pH levels less than 4.0 such as steel industry wastewaters and acid mine drainage.

FINDING A LIMESTONE SUPPLIER

Limestone is a basic construction material used nationwide. It is also used as a raw material in a variety of manufactured products, including cement and ice cream. Composed mainly of calcium carbonate ($CaCO_3$), almost all natural limestones contain varying amounts of magnesium.

Many rich deposits of limestone exist throughout the United States, with numerous operating limestone quarries. Millions of tons of limestone are quarried each year, primarily for agricultural, environmental, and industrial applications. Even if the surface water to be limed is located in a watershed that does not contain sufficient limestone rock, a limestone quarry probably exists within 160 km (100 miles) of the waterbody.

Finely ground limestone (10–30 μm) is the preferred material for most surface water neutralization techniques, although the finer the grinding grade, the higher the cost of the material. Nevertheless, most of the hundreds of U.S. limestone quarries do not grind limestone into the finer particle sizes recommended for some water applications. However, some of the larger eastern U.S. vendors, including Omya, Inc. (61 Main Street, Proctor, Vermont 05765) and Pfizer Chemical Company (235 East 42nd Street, New York, New York 10017), provide a variety of grinding grades, including the finer grades of limestone for the paper, paint, food, asphalt, metal alloy, and other industries. Many other vendors supply agricultural limestone, but it is generally too coarse for most liming techniques. However, some stream dosing devices such as diversion wells and rotary drums use coarse material that is widely available.

SELECTING THE MATERIAL

When selecting a liming agent, three special properties of commercial limestone must be considered. One is the product's calcium content, expressed as percent $CaCO_3$ or CaO; another is the size of the limestone particles, commonly referred to as the grinding grade; and a third is the presence of contaminants, such as nutrients or toxic metals. Each of these properties affects treatment results.

Calcium Content

In simplest terms, the calcium content of the limestone is directly proportional to its capacity to neutralize surface water acidity. For example, 1,000 tonnes of a limestone product containing 90 percent $CaCO_3$ yields an acid neutralizing effect equivalent to 900 tonnes, while material with 75 percent $CaCO_3$ has an equivalent of 750 tonnes. The limestone material should be at least 70 percent $CaCO_3$ by weight, and 90 percent is preferable.

Particle Size

Limestone materials are available in a variety of particle size ranges (also referred to as grinding grades). Each liming technique uses different grinding grades. Smaller particles of limestone have a greater surface area per unit weight than larger particles. The smaller particles, therefore, produce greater reactivity.

The average particle size for limestone materials may be defined in several ways, based on diameter, number of particles, particle surface area, and particle volume. The average particle size used throughout this book is based on the particle surface area and is referred to as the surface area-weighted mean particle diameter. This term is the average diameter weighted to an individual particle's surface area. The particle size is commonly measured with either wet sieving or Coulter counter techniques.

In general, the coarser agricultural limestone (average particle size greater than 0.5 mm) is not recommended for treatment of lakes. Although this material is often available locally, it will not result in cost-effective or labor-efficient treatment in most surface waters. The larger particles in conventional agricultural limestone result in poor dissolution efficiency. For example, only 15 percent of particles greater than 0.2 mm dissolve in mildly acidic waters. As a result, the amount of limestone that will remain in the sediment after application of this size material is often far more than is needed to neutralize sediment acidity or that can dissolve over time during reacidification. Moreover, the excess limestone in the sediment may become coated with iron, soil particles, or other materials, and it no longer reacts with surface water acidity.

The distribution of particle sizes for different commercially available limestone powders is shown in Figure 5-1. These curves are based on the results of wet sieving. The vendor will often provide information on particle size distribution that includes this type of distribution curve and, frequently, additional data on the bulk density of the material.

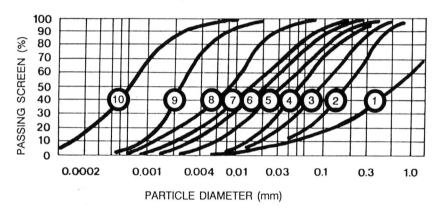

Curve No.	Classification Description
1-2	0-3 mm agricultural limestone
3-4	0-2 mm agricultural limestone
5	0-1 mm agricultural limestone
6	0-0.5 mm powdered limestone
7	0-0.25 mm powdered limestone
8	0-0.05 mm fine powdered limestone
9	0-0.005 mm chalk filler
10	0-0.002 mm marble filler

Figure 5-1.—Particle size distribution curves for different limestone powders (Source: Olem, 1991).

In general, average particle sizes between 10 and 30 μm are preferred for most lake treatments because this material

- provides adequate neutralization of the water column to >5 mg/L as $CaCO_3$ (>100 μeq/L ANC),

- satisfies sediment neutralization requirements,

- allows for residual sediment neutralization to slow the process of reacidification, and

- results in lower unit cost because of more efficient dissolution.

The preferred particle size for streams depends on the particular device used for stream treatment. Recommended particle sizes for different devices are discussed in the section on stream application methods.

Presence of Toxic or Unwanted Constituents

Maximum acceptable levels for chemical constituents in limestone materials are shown in Table 5-1. Generally, the dolomitic limestones are not as effective as the high calcium materials; therefore, material containing no more than 5 percent $MgCO_3$ by weight is preferred. $MgCO_3$ has a lower neutralizing capacity than $CaCO_3$ and usually dissolves more slowly.

Table 5-1.—Acceptable maximum levels for chemical constituents in limestone materials used for neutralization.

PARAMETER/ELEMENT	PERCENT BY WEIGHT
Magnesium carbonate ($MgCO_3$)	5
Organic materials[a]	5
Phosphorus (P)	0.1
Aluminum[b] (Al)	1
Manganese (Mn)	1
Lead (Pb)	0.1
Mercury (Hg)	0.0005

[a] Loss at 200°C (390°F)
[b] Most often found in the fairly inert form Al_2O_3

If the $MgCO_3$ is about 5 percent by weight, then $CaCO_3$ typically is greater than 90 percent, and the remaining materials include primarily acid insoluble compounds such as silicon dioxide. In situations where the material may be used in foods, the product must meet the stringent maximum contaminant levels required by the U.S. Food and Drug Administration Food Chemical Codex.

Limestone Quality

Limestone vendors generally adhere to a quality assurance and quality control (QA/QC) program that usually consists of chemical and physical monitoring of the limestone material. Results from vendors' analytical laboratory analyses are often available for review and provide important information to determine whether a material meets selected criteria.

The data that should be obtained from the vendor are

- chemical analyses required by the Food and Drug Administration for food-grade products as outlined in the Food Chemical Codex,

- other chemical data on total phosphorus, total nitrogen, and total organic carbon, and

- physical properties such as particle size distribution and bulk density.

These data should be obtained from the supplier with each new lake treatment operation and from 10 percent of the shipments of bulk or bagged limestone delivered to a continuous stream treatment site. A sample of each batch of material delivered to the site should also be archived.

Dispersant Materials

Chemical dispersants, such as sodium polyacrylates and guar gum, can be used in aerial applications to keep the limestone slurry in suspension during transport. Dispersants allow a more concentrated slurry (thereby reducing transportation costs) and allow longer storage.

Dispersants can also be used in stream doser slurry storage tanks to keep the particles dispersed before application. Dispersants are usually unnecessary for lake applications in which a slurry box or barge is used. Dispersants may also increase dissolution efficiency by reducing the aggregation of limestone particles after application to streams and lakes.

A review of the literature found no reports of negative effects from use of chemical dispersants. If a particular application will benefit from the use of dispersants, contact the supplier of limestone materials to get them. If the supplier does not market dispersants, the company should be able provide names of suppliers.

TRANSPORTING AND HANDLING THE LIMESTONE

Transportation Considerations

Bulk trucks—primarily vehicles hooked up to a trailer and capable of hauling loads of 30 to 35 tonnes of limestone—are the most commonly used means of transporting material to a liming site where more than one truckload of limestone is needed. Limestone vendors have fleets of such trucks and generally charge haulage fees based on the distance traveled. More information on costs is included in a later section.

Trucks containing bags of limestone often transport one or two truckloads or less as an alternative to bulk trucks. Good roads for transporting material to the treatment site are often a major consideration in determining the exact method of transport. Typically, sites where liming is needed are in remote areas and roads are either inadequate for heavy equipment or generally lacking. Roads used for limestone delivery must be designed to carry heavy equipment, and bridges must be able to sustain heavy loads. Sites that are not accessible to heavy trucks may need light trucks carrying smaller loads. Alternately, trucks can deliver material to an accessible staging area for aerial application.

These transportation considerations also apply for limestone trucked to staging areas for aerial applications. Generally, however, the staging area is set in a more accessible area than the site where the liming will take place. The ground at the site must also be able to support heavy vehicles.

Material Handling

Transferring liming material from a vehicle to the treatment vessel or storage silo is often a difficult and time-consuming procedure. Small amounts of limestone transported in bags can be unloaded by hand. For larger bagged loads of limestone, a forklift can unload the bags. When bulk limestone is delivered to a treatment site, the truck can usually unload the material directly into the dispersing or storage unit. Sometimes, pneumatic equipment or construction vehicles (front-end loaders, for example) load the material.

The total costs of various types of limestone materials may differ from their cost per unit of weight. Table 5-2 shows the delivered costs of the more commonly used materials (limestone gravel, limestone powder, and food-grade powdered and slurried limestones).

Table 5-2.—Typical costs, including transportation, for selected limestone materials.[a]

MATERIAL	COMMERCIAL CLASSIFICATION (mm)	MEAN DIAMETER[b]	COST/ TONNE[c]
Limestone gravel	20–50	38 mm	35
Limestone powder (Curve 5)[d]	0–1	18 µm	68
Limestone powder (Curve 6)[d]	0–0.5	14 µm	74
Limestone powder (Curve 7)[d]	0–0.2	12 µm	80
Limestone powder (Curve 8)[d]	0–0.044	7.5 µm	90
Limestone slurry (Curve 9)[d]	0–0.005	5.5 µm	135
Limestone slurry (Curve 10)[d]	0–0.002	0.7 µm	145

[a] Assumes transport distance of 80 km (50 miles) within northeastern United States. Does not include provision to download material at site with a forklift or other device
[b] Surface area-weighted mean
[c] To convert to $/ton, multiply by 0.907
[d] Refer to Figure 5.1 for particle size distribution curves

Another important consideration for material handling is the time of year of the treatment. It may be advantageous to avoid periods when it is likely that heavy rains will cause muddy conditions that may affect material handling and transportation. Of course this factor should be weighed along with ecological considerations for the time of year.

Storage Considerations

Limestone is hydroscopic—meaning it has a tendency to absorb moisture. Wet material clumps together and clogs distribution equipment. The finer the grade of limestone, the more susceptible it is to absorbing moisture. Therefore, the material must be protected from moisture when it is stored.

For short-term storage at lakeshores, plastic tarps can protect bag limestone. Stream liming devices generally use a storage silo fabricated from metal or other strong and weather-resistant materials in which limestone can be stored.

COSTS OF LIMESTONE MATERIALS

The total costs of various types of limestone materials may differ from their cost per unit of weight. Table 5-2 shows the delivered costs of the more commonly used materials (limestone gravel, agricultural limestone, and food-grade powdered limestones). The transportation charge based on the distance from the material source to the liming site is a major factor in the total price of the materials. The material cost in Table 5-2 assumes a transport distance of 80 km (50 miles) within the northeastern United States. Transportation increases the average cost of the neutralizing materials $28 per tonne ($25/ton).

Limestone costs vary widely. The tabulation in Table 5-2 does not incorporate the neutralizing efficiency of the material or the variations in efficiency of application. These factors must be included in calculating the cost of an application.

Limestone aggregate has the lowest delivered cost per ton among the selected materials. This material, however, is used in only a few limited types of stream treatment techniques such as distribution by rotary drums. The equipment costs are higher for devices that use aggregate instead of ground limestone.

Limestone powders with a mean diameter of 7.5 to 14 μm are the most commonly used neutralizing materials in both lake and stream applications. Delivered costs range from $74 to $90 per tonne ($67 to $82 per ton)—much less than the more finely ground limestones. Shallow lakes often benefit from use of finer particle sizes (10 to 20 μm) because much of the limestone will dissolve prior to reaching sediments. Larger grinding grades are often specified for deep lakes and certain stream liming devices. Using unit costs, such as cost per year or per unit of lake surface area, may provide more accurate cost estimates for an application than using material cost per unit of weight.

For example, the finer limestone powders (mean diameter 0.7 to 5.5 µm) cost much more than coarse grained powders (mean diameter 7.5 to 18 µm), but because of their greater reactivity, the quantity required may be reduced.

Bagged material typically costs about 10 percent more than bulk material. Also, a premium is charged for transporting less than a full truckload. Moreover, transportation costs do not include unloading of bagged material at the site (normally an additional $100 to $200).

Limestone slurries are sometimes dispensed from automated dosers or dropped from helicopters. Some managers apply powders instead of slurries, offsetting the initially high costs of powders in part by eliminating the need for the labor and equipment involved in a slurrying operation (typically 30 percent of the material cost).

SELECTING LAKE, STREAM, OR CATCHMENT LIMING

A variety of methods exist for distributing limestone to aquatic systems. It can be applied directly to lakes and streams, or it can be added to upstream land or wetland areas in the watershed.

Equipment required for a specific application is determined largely by application needs, cost, and accessibility. Direct limestone application into a lake is the most commonly used method and is generally the most cost effective. Stream liming generally requires equipment that is costly and expensive to maintain. When lake or stream liming is not economical, as when a lake has a short water retention time and would need to be treated more often than once a year, liming soils and wetland areas is receiving increased attention as an alternative.

Figure 5-2 presents a decisionmaking strategy for determining whether to apply limestone to a lake, stream, or watershed. First, the lake or stream to be managed must be evaluated to determine water retention time (for lakes), flow (for streams), watershed characteristics, and other information about the waterbody and its surroundings. The decision whether to lime the lake or stream directly is based on the answers to questions about the particular conditions at the site, including the water retention time and whether an upstream lake can be limed. The next sections discuss these factors and the various application methods available.

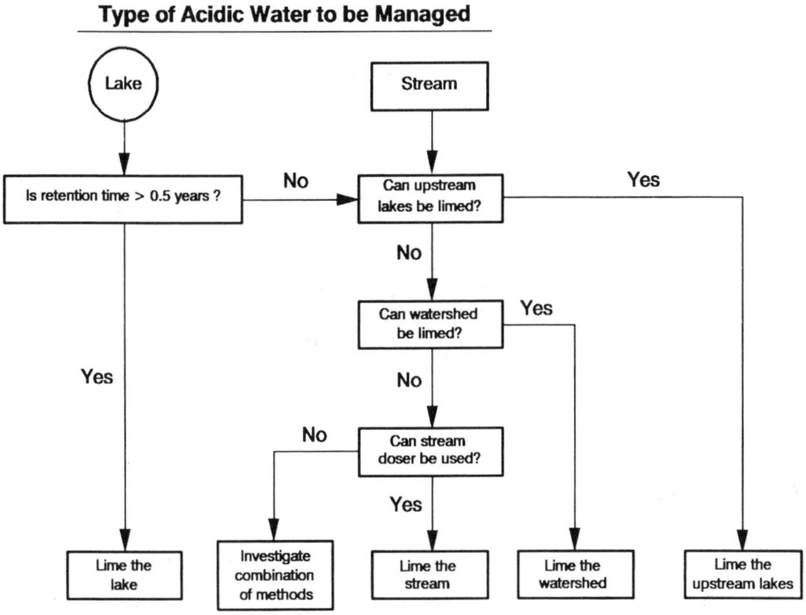

Type of Acidic Water to be Managed

Figure 5-2.—Decision strategy for determining whether to lime the lake, stream, or watershed.

The advantages and disadvantages of the various application methods for lakes, streams, and watersheds are summarized in Table 5-3. Only a handful of companies specialize in applying limestone to lakes and designing and constructing treatment devices for streams. Sweetwater Technology Corporation (P.O. Box 3370, Palmer, Pennsylvania 18043), which conducts lake liming projects in the United States, has installed a few stream liming devices. A number of stream liming devices are available in Sweden and Norway, although they are not readily accessible. For example, Boxholmkonsult AB (Box 86, 590 10 Boxholm, Sweden) and Cementa Movab AB (Box 30022, S-200 61 Malmo, Sweden) specialize in the design and installation of the stream dosers described in stream application methods later in this chapter.

Table 5-3.—Advantages and disadvantages of lake application methods.

TECHNIQUE	RELATIVE USE	ADVANTAGES	DISADVANTAGES
Boat	Common	Simple, effective, and accurate. Allows different distribution options. Commercial applicators exist.	Not very practical for remote sites. Cannot treat ice-covered lakes.
Truck/ Tractor	Occasional	Can distribute materials more quickly than boat method. Can treat ice-covered lakes.	Not practical where road access is limited. Able to treat only small lake from shore or ice-covered lake. Ice application may be dangerous.
Aircraft	Occasional	Allows applications at remote sites. Less labor-intensive once loaded. Distribution of materials is quicker than by boat or truck.	Expensive

LAKE LIMING

Direct liming is the preferred method for lakes with a water retention time greater than one half year (Simonin et al. 1990; Olem, 1991). If the retention time is less, the feasibility of liming an upstream lake that may have a longer retention time should be considered instead. If this option is not possible, consider the methods for liming watersheds and streams described in the next sections.

When direct lake liming is the logical choice after considering the other alternatives, the next decisions are when to lime, where to place the material, how much to add, and what type of application method to use.

Timing of Liming

Periods of high influxes of additional acid, such as spring snowmelt and fall rains that coincide with reproductive activity, create critical times in lakes that support existing fish populations. Applications should be scheduled to precede critical events if possible.

Application during spring overturn takes advantage of lake circulation patterns to enhance the mixing and distribution of limestone. However, this period may be too late to prevent the mortality of embryos and fry of numerous fish species that occurs during spring snowmelt. Liming for protection from melting snow may best be done during or shortly before fall overturn, especially if fish are to be stocked soon after liming.

Immediately stocking fish, however, is not recommended unless water quality analyses reveal that conditions are safe for the particular fish species. The relationship between the timing of liming and fish stocking should therefore be given careful consideration before applying the limestone. Fish stocking considerations are discussed in Chapter 6.

Location of Liming

Distribution of limestone over an entire lake may be the ideal way to guarantee neutralization of the waterbody. However, this is not always practical or warranted because of funding, time, and technical factors. If application of limestone to the entire lake is not feasible, it may be possible to effectively neutralize the waterbody by distributing the material over the deepest portion of the lake, thus increasing the amount of time for the larger particles to react while they pass through the water column. This distribution method is probably best suited for deep lakes (greater than 8 m [26 ft]) that have relatively rapid flushing rates (less than 2 years retention time).

The primary consideration in deciding which parts of the lake to lime is whether the choice most effectively benefits the biota. For example, liming only shallow littoral zones may be desirable if protection of early life stages is critical.

Calculating the Dosage

The complexity of dose calculation methods varies widely (Sverdrup and Warfvinge, 1988; Olem, 1991). During the early years of liming, dose calculations were often only rough approximations based on acid-base titrations. Sometimes, the limestone dose added to lakes was too small to raise pH and alkalinity to sufficient levels. In other cases, too much limestone was added, resulting in accumulation of unused limestone in the sediments. Some extra limestone in the sediments can prolong neutralized conditions, but adding too much limestone wastes time, money, and natural resources.

Now, rule-of-thumb calculations based on many years of liming experience exist to determine limestone doses. Also, sophisticated models may be used that incorporate many of the processes important in selecting the proper dose. The amount of limestone required varies according to the following factors:

- where the material is being applied,
- type of material used, such as the particle size and calcium content,
- material dissolution rate,
- water temperature,
- volume of the waterbody,
- flushing rate of the waterbody,
- acidity of the water,
- acidity and makeup of associated sediments,
- amount of organic materials in the water,
- acid input from incoming water, and
- acid input from atmospheric deposition.

Dose calculation methods have been developed that are based on these conditions. From a practical standpoint, however, it is possible to use an empirical method that takes these factors into account by incorporating the results of numerous limestone applications. A lime dose model that requires minimum data inputs appears in the box on page 61. It uses two graphs (Figs. 5-3 and 5-4) and three simple equations to compute the required dose.

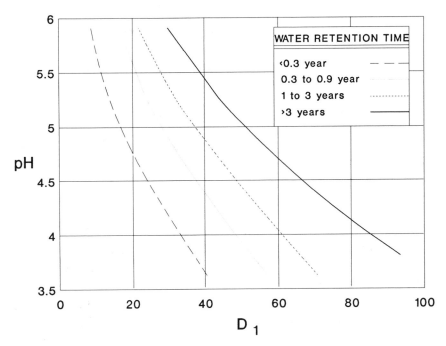

Figure 5-3.—Step 1 of lime dose model: calculation of D_1.

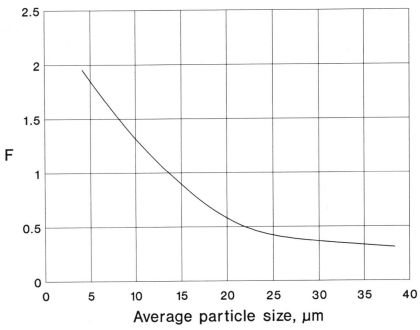

Figure 5-4.—Step 3 of lime dose model: calculation of F.

The procedure shown in the box on page 61 is an empirical model for whole-lake treatment based primarily on the thousands of liming projects conducted in Sweden. A more complete discussion of available models is presented by Olem (1991).

Calculating Reacidification Time for Lakes

Calculation of reacidification rates is important to provide an estimate of when to re-treat a particular lake to maintain the prescribed water quality and fisheries. This is essential information for determining whether a particular application is economically feasible. Unfortunately, estimates may be uncertain because of variations in the data inputs such as climatic conditions that differ from past records. Therefore, frequent water sampling is a good supplemental procedure for determining when to re-treat a lake.

Several calculations methods can predict the rate of reacidification of a lake treated with limestone. The calculations for determining the reacidification rate use data similar to those for calculating lake chemical dose, such as the characteristics of the lake water and sediments, the hydrologic and morphometric conditions in the watershed, the quantity and quality of precipitation in the watershed, where the limestone is to be applied, and the characteristics of the limestone material.

Models for predicting reacidification rates incorporate many aspects of these factors. DePinto et al. (1987), Sverdrup and Warfvinge (1988), and Olem (1991) provide information on available models. The simplest models incorporate important factors through empirical relations. A simple model based on an empirical relation with lake retention time is presented here in Figure 5-5. It shows the duration of neutralized conditions (represented as pH >6.0 or alkalinity >2.5 mg/L as $CaCO_3$ (>50 µeq/L)) for lake water retention times up to 10 years.

Lake Application Methods

Three basic techniques for applying limestone to lakes include boat or barge, truck or tractor, and helicopter or fixed-wing aircraft.

OLEM LIME DOSE MODEL

Data required:

- pH before liming
- lake retention time (yr)
- lake volume (m^3)
- average limestone particle size (μm)
- calcium content of limestone (in percent CaO, obtained from manufacturer)

The model assumes water quality targets described in Chapter 3 (pH 6.5 and ANC 5 mg/L as $CaCO_3$ (100 μeq/L)).

Calculation steps:

STEP 1: Estimate D_1

Using the lake water pH before liming and the water retention time, D_1 is estimated using Figure 5-3.

STEP 2: Modify the dose for limestone calcium content

The calcium content of the limestone, C (expressed as percent CaO), is entered into Equation 5-1.

$$D_2 = D_1 \times 60/C \qquad \text{5-1}$$

where D_2 is the dose factor adjusted for calcium content,
D_1 is the dose factor with no adjustments for limestone characteristics estimated in Step 1 (Figure 5-3), and
C is the percent calcium as CaO.

STEP 3: Modify the dose for limestone particle size

The average particle size of the limestone is used in Figure 5-4 to determine the dissolution factor, F. This factor is entered into Equation 5-2.

$$D_3 = D_2/F \qquad \text{5-2}$$

where D_3 is the dose adjusted for the limestone particle size and calcium content in g/m^3,
D_2 is the dose factor adjusted for calcium content, and
F is the dissolution factor estimated from Figure 5-4.

STEP 4: Calculate the dose in tonnes

The required dose for limestone with a calcium content, C, a mean particle size, P, is calculated using Equation 5.3.

$$D = D_3 \times V/1,000,000 \qquad \text{5-3}$$

where D is the dose in tonnes,
D_3 is the dose in g/m^3, and
V is the lake volume in m^3 (see Chapter for a discussion of lake volume calculation methods).

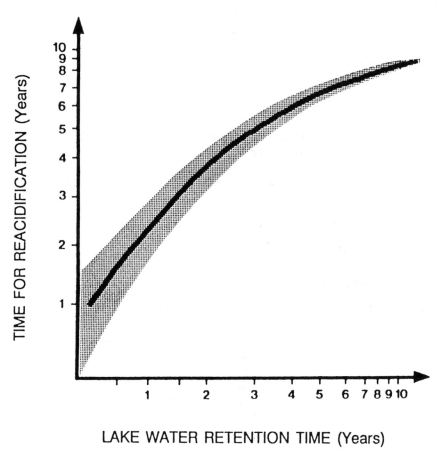

LAKE WATER RETENTION TIME (Years)

Figure 5-5.—Prediction of time for reacidification: the duration of neutralized conditions based on lake water retention time. Shaded area represents estimated prediction error (Source: Olem, 1991).

Boat or Barge

Application from a boat is probably the least expensive method for accessible lakes and is now the most commonly used method.

For relatively small applications, the technique usually involves slurrying the limestone on board the boat using a tank filled with lake water. Bags of limestone are emptied into the slurry box and the slurry is pumped from the tank and sprayed into the lake (Fig. 5-6). The technique is simple and inexpensive, but it is labor-intensive and slow. The maximum applica-

Figure 5-6.—Spreading of limestone from slurry box on a boat. (Photograph courtesy of Living Lakes, Inc.)

tion rate is 2 tonnes per hour (2.2 tons per hour) with a single boat.

For larger applications, pontoon boats or barges are equipped with large, pressurized storage tanks, compressed air blower systems, slurrying units, generators, and pumps to supply lake water to the slurrying units. Pressurized tank trucks pneumatically deliver dry powdered limestone from the shore into storage compartments on board the boat. Lake water is pumped into a small slurry tank, mixed with limestone, and sprayed onto the lake surface. The only commercial system currently operating in the United States was designed and built by Sweetwater Technology Corporation (Fig. 5-7).

Truck or Tractor

Land vehicles such as trucks and tractors are used to transport and distribute liming materials to lakes, particularly to spread powdered limestone on ice in northern climates.

Figure 5-7.—Commercial barge-pressure tank application of limestone in the United States. (Photograph courtesy of Sweetwater Technology Corporation.)

Blowing dry material from shore onto ice or water is not a viable method for lakes because of poor control of dosage and dissolution of limestone. Distributing slurry from a truck or tractor may be effective for relatively small lakes where there is adequate access to the lake surface from the shoreline.

Helicopter or Fixed-wing Aircraft

Application of limestone by aircraft is usually the most efficient and cost-effective method for remote waters. This method, however, is more costly per unit of material applied than using application by boat. Use of helicopters for liming is more common, more accurate, and generally less costly than fixed-wing aircraft. Helicopters spray dry powdered or slurried limestone by slowly traversing the lake target zones (Fig. 5-8). Commercial systems set up specifically for application of limestone by helicopter are currently available in Sweden but not in the United States.

In the helicopter method for releasing dry limestone powder, a filled bucket is transported to the target water. The helicopter maintains an altitude of 50-230 m (150-750 ft) over the waterbody, while a door in the bottom of the bucket opens

Figure 5-8.—Helicopter application of limestone.

electronically, releasing the load. More than one bucket is used to maximize efficiency, with buckets alternately filled at a staging area. The dry powder technique may inaccurately distribute material during windy conditions. It can also release fine particles that may obscure the pilot's vision and possibly damage the helicopter rotor blades. Prudent pilots carefully consider wind speed and direction before liming.

The slurry technique is often a better aerial application method because it can distribute the material more accurately, even in windy conditions. In this method, a helicopter is fitted with a storage tank that is filled with slurried limestone from a tank truck. Using a spray nozzle similar to that used for hydroseeding of clear-cut or strip-mined areas, a helicopter with a 1-tonne storage tank can apply 5 to 7 tonnes per hour if the target lake is within 10 km (6 miles) of the staging area.

Fixed-wing aircraft have had limited use in lake liming, primarily because the aircraft is not as maneuverable as a helicopter and a runway nearby is needed as a staging area.

Still, commercial equipment is available and experience already exists in the application of water in fire-fighting operations and agricultural chemicals in farming. The technique may be most useful for larger lakes (> 50 ha [> 120 ac]).

Costs of Lake Liming

Few studies have comprehensively evaluated the costs of liming techniques. In them, the expenses cited typically include costs of materials, transportation, and application. The costs of liming materials, including transportation, were presented earlier in this chapter. The choice of liming material and the application technique will affect both the resource requirements and costs. Costs associated with the various lake liming techniques described previously are included here. The data should be used with caution, however, because of the wide variability in costs for different situations, such as available materials, application techniques, and site accessibility.

Cost data presented here were often obtained from research and demonstration liming programs. In most of these studies cost optimization was not a major consideration. For example, because of the desire for precise data on application quantities, the application may have taken three times as long as an ordinary application, increasing the total costs.

Another complication in comparing costs is that candidate sites for liming are sometimes far from the source of the base material. For example, limestone quarries are rarely found near lakes in areas of granite bedrock. This is not a universal problem, however. Limestone vendors in the United States are ordinarily within 50 to 80 km (30 to 50 miles) of candidate sites, unlike parts of Sweden and Norway. Another cost complication is the unavailability of labor near the typical acidic lake, which is often remote and inaccessible.

Although it is helpful to know the initial costs for a particular treatment technique, the actual cost is amortized annually and includes successive treatments. Also, other tasks are not usually considered that are part of the costing in an aquatic liming program, such as the costs of site assessment, obtaining permits, characterization, monitoring, and administration. These costs can sometimes add substantially to the material and application costs.

Cost ranges of the various liming techniques for lakes are summarized in Table 5-4 and include costs for materials, transportation, and application. The costs are based on actual liming applications after 1970 and are converted to 1991 U.S. dollars assuming 5 percent per year inflation. Costs may vary because of accessibility, use of donated labor, and cost variations in different locations. Costs of site selection, pre- and post-treatment monitoring, obtaining permits, report preparation, and administration are not included because of the wide variations in these costs among liming projects.

Table 5-4.—Typical cost ranges for lake liming application methods.[a]

APPLICATION METHOD	COST/TONNE[c]			COST/HA[d]		+
	MATE-RIALS[b]	APPLICA-TION	TOTAL	MATE-RIALS	APPLICA-TION	TOTAL
Boat: Manual dry spreading	$25–100	$25–250	$50–350	$25–100	$25–250	$50–350
Boat: Slurry box spreading	25–100	25–250	50–350	25–100	25–250	50–350
Boat: Barge and pressure tank	25–100	25–250	50–350	25–100	25–250	50–350
Truck or tractor	25–100	50–500	75–600	25–100	50–500	75–600
Helicopter	50–150	150–500	200–650	50–150	150–500	200–650
Fixed-wing aircraft	50–150	200–500	250–650	50–150	200–500	250–650

[a] Costs converted to 1991 U.S. dollars assuming 5% per year inflation. Costs vary because of accessibility, use of donated labor, and cost variations in different locations, including charges in foreign currency exchange rates. Costs of site selection, pre- and post-treatment monitoring, permits, report preparation, and administration are not included.
[b] Includes material transport costs.
[c] To convert to cost/ton, multiply by 0.907.
[d] To convert to cost/acre, multiply by 0.404.

STREAM LIMING

It may not be necessary to lime acidic streams directly. First, you should determine whether an upstream lake can be limed. Lake liming is generally simpler and, unlike continuous stream dosing, is performed once yearly or even less frequently. If upstream lake liming is not possible, then explore the feasibility of watershed liming. Neutralized conditions in downstream waters may last even longer with watershed liming than liming an upstream lake. However, experience with watershed liming is limited, and it should be considered only after a thorough investigation of site conditions and the results of previous watershed applications (see next section).

If direct stream liming is the logical choice after considering the other alternatives, decisions include when to lime, how much lime to add, what type of device to use, and where to place it.

Timing of Liming

For streams, spring storms may be a particular cause for concern because of the exposure of fish eggs and larvae to episodes of high acidity. It may be desirable to lime streams only during the fish spawning season or for storms of a certain magnitude (for example, when triggered by stream stage height or flow changes).

Location of Liming

The most important concern for location of stream liming devices is providing a zone where metals may precipitate by placing the devices at some point above the target area. Of course, not all candidate streams have concentrations of metals that precipitate and negatively affect biota. In these cases, a secondary concern is selecting a point on the stream where the limestone mix and dissolve sufficiently. Water-powered devices need considerations of water pressure requirements and, often, the need for some type of water regulation device, such as a small dam to create a specified drop in stream head. How these concerns are handled depends on the specific conditions along the stream reach to be treated and the device selected for applying the limestone.

Calculating the Dosage

Many of the same factors that determine the dose required for lakes are applicable to streams. The limestone dose for streams varies according to the following factors:

- where the material is being applied,
- type of material used, such as the particle size and calcium content,
- material dissolution rate,
- stream flow,
- water temperature,
- acidity of the water,
- acidity and makeup of associated sediments,
- amount of organic materials in the water,
- acid input from incoming water, and
- acid input from atmospheric deposition.

Because of the many possible types of stream liming devices, dose calculations for streams can be obtained by the vendors who provide the equipment. Approximate quantities of powdered limestone needed for electrical and water-powered stream dosers is presented in Figure 5-9. The dose is estimated from the stream pH before liming. This information can be used for initial planning purposes to determine the amount of limestone to deliver to the stream site. For stream dosers, the amount of limestone added at any time typically varies according to a surrogate parameter, such as stream stage height. Larger, more sophisticated systems sometimes use a relationship between pH or conductivity and dosage.

Stream Application Methods

The design and operation of stream liming devices rely in part on experience gained with systems for treating acid mine drainage, industrial water and wastewater treatment, and related industrial operations. The technology has advanced

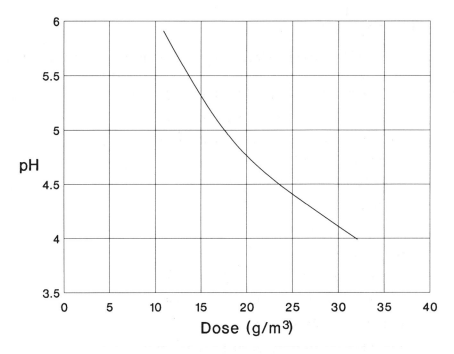

**Figure 5-9.—Recommended doses for liming streams using powdered lime-
stone. Doses are based on limestone with 50 percent CaO.**

rapidly over the last few years, particularly in Sweden and Norway.

Techniques tried for stream liming include several types of
dosers (any mechanical device that releases powdered or slurried material into streams), diversion wells, limestone barriers,
manual application of limestone fines, and rotary drums. The
most common devices, discussed here, are the electric-powered
doser, water-powered doser, diversion well, and rotary drum.
Characteristics of these systems are summarized in Table 5-5.
See Olem (1991) for a discussion of other stream liming devices.

Electric-powered Doser

An electric-powered doser distributes either dry powdered
or slurried limestone directly into the stream. A dry-powder
doser consists of a limestone storage bin, a feeder screw
operated by battery or line voltage, an automated dose control
mechanism, and a distribution pipe that dispenses the powder

Table 5-5.—Comparison of techniques used in neutralizing acidic streams.

FACTOR	LINE VOLTAGE ELECTRICAL DOSERS	BATTERY-POWERED DOSER	WATER-POWERED DOSER	DIVERSION WELL	ROTARY DRUM
Maximum flow (m^3/sec)[a]	25	8	3	1	1.5
Stream head required (m)[b]	0	0	1.0	> 1.3	> 1.5
Moving parts	Yes	Yes	Yes	No	Yes
Electrical power required	120/240 Volt AC	12 Volt DC	None	None	None
Type of neutralizing agent normally used	Slurried limestone powder	Dry limestone powder	Dry limestone powder	Crushed limestone	Limestone aggregate
Other limitations	Sensitive to thunderstorms/power failures. Wet-slurry intake piping may become clogged.	Sensitive to freezing (below -5°C)	Sensitive to freezing (below -5°C)	May not perform well during low or high flows or for highly acidic systems.	Sensitive to freezing (below -5°C); generally higher initial cost.

[a] To convert to ft^3/sec, multiply by 35.3.
[b] To convert to ft, multiply by 3.28.

to the stream. Typically, limestone feed is automatically regulated by the water level, a surrogate for water flow. As the water level fluctuates, more or less limestone is transported from the storage bin by the feeder screw to a conveyor belt. This belt, located within a pipe, transfers the powder to the point of distribution above the center of the stream.

Some dosers slurry the material with stream water on site before dispensing it (Fig. 5-10). Water is pumped from the stream to the slurry tank where it is mixed with the limestone powder. The resultant slurry is pumped to the dispersion well in the stream bed, mixed with additional stream water, and dispersed just below the stream surface.

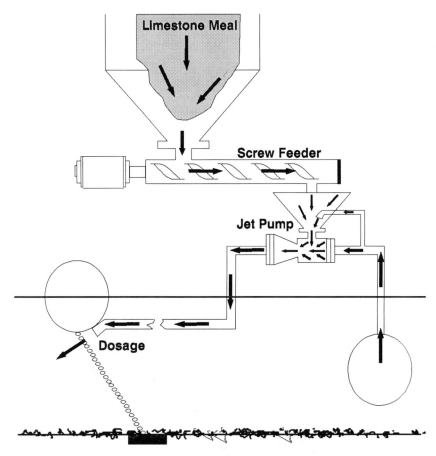

Figure 5-10.—Diagram of an electrically powered doser dispensing limestone as a wet slurry (Source: Cementa Movab, Malmo, Sweden).

Some dosers store and dispense a commercially prepared slurry of extremely fine limestone powder into streams (Fig. 5-11). The slurry is pumped directly from a storage tank to a Y connection, where it is diluted with water before dispersal. The slurry in the tank is stirred frequently to avoid sedimentation.

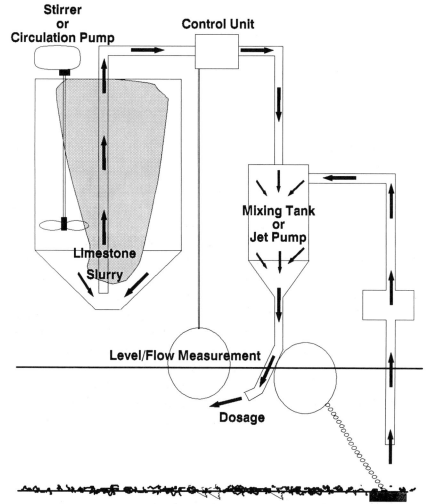

Figure 5-11.—Diagram of an electrically powered doser storing commercially prepared limestone slurry (Source: Cementa Movab, Malmo, Sweden).

The most common dosers are operated by line voltage, and these are recommended where electrical power is available. Battery-powered dosers are sometimes used in remote locations where line power is unavailable. Because of power limitations, these systems are typically smaller installations with storage bins filled manually with bagged limestone.

Other common features of electric-powered dosers include a tapered storage bin to prevent powder buildup on its walls and a hammer, or similar device, that periodically strikes (or vibrates) the wall to loosen any compacted material.

Water-powered Doser

In water-powered dosers, regulated streamflow controls the dispersion of limestone. The dosers are installed either in or adjacent to the stream. They operate using either bucket feed mechanisms or an apparatus similar to a paddle wheel that turns an auger.

The bucket feed system is widely used in Sweden. The Boxholm doser consists of a storage bin with a small opening in the bottom, a conveyor belt, and a rocker arm with a hammer mechanism at one end and a tipping bucket at the other end (Fig. 5-12). A small portion of the stream flow, manually regulated by a valve, is diverted through a pipe or trough into the bucket attached to the rocker arm. As the bucket fills with water, a continuous flow of limestone is gravity-fed to a conveyor belt located a few centimeters below the mouth of the limestone storage bin. When the bucket is filled with the stream water it tips over, causing the conveyor belt to spill limestone into the bucket. The movement of the bucket also causes the hammer mechanism to strike the side of the storage bin, loosening any powder that has lodged onto the sides of the bin.

Diversion Well

A diversion well is a container filled with crushed limestone—usually 6 to 8 mm in diameter—and located along a stream bank or in the bottom of a streambed (Fig. 5-13). The device receives all of its energy from the water, which is diverted from the stream through a tube that opens at the bottom of the well. The diverted water flows upward through the

Figure 5-12.—Drawing of the "Boxholm" water powered doser (Source: Box-holm Konsult AB, Boxholm, Sweden).

crushed limestone, agitating the material and creating a fluidized bed that results in abrasion and dissolution of small alkaline particles. The neutralized water then escapes over the lip of the well or through an outlet pipe. Increasing the diameter of the top of the well decreases the fluid velocity at

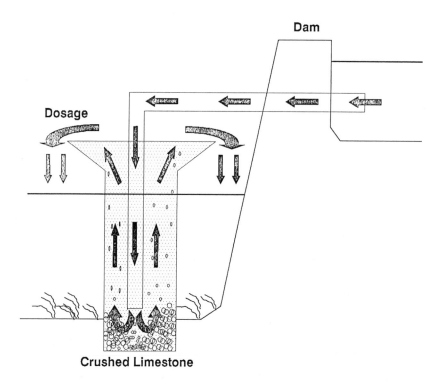

Crushed Limestone

Figure 5-13.—Diagram of a diversion well (Source: Cementa Movab, Malmo, Sweden).

the outlet, and undissolved particles sink back into the well rather than being entrained in the outflow.

The first step in the design of this type of lime delivery system is to determine the amount of limestone required in a daily dose, based on the flow rate and neutralization demand of the stream. Daily dose is calculated by titrating a sample of stream water with limestone and multiplying the result by the daily flow rate. To calculate the total amount of limestone needed in the well, divide the amount of limestone required in a daily dose by the percent of the total amount of limestone dissolved in the well per day. The amount dissolved in the well per day can be determined by using the inlet velocity and the diameter of the well to compute the linear upward velocity of the water in the well. Example calculations are presented in Fraser et al. (1985).

Sufficient pressure to fluidize the limestone bed requires a hydrological head of 1.3 m (4 ft). The process typically results in use of about 80 to 90 percent of the limestone added. Treatment of flows up to about 1.0 m^3/sec (35 ft^3/sec) is possible using two or more wells in parallel. The system is generally capable of increasing the pH of stream water a maximum of 1 or 2 units and may not perform well at extremely high or low flows.

Rotary Drum

Rotary drums are cylindrical containers filled with limestone aggregate and powered by water diverted from the stream and directed across a sluiceway. Openings in the bottom of the sluice are located directly above each drum, which serves as the hub for waterwheel style blades on the exteriors of the drums. As water falls through the sluice openings, the drums rotate.

A hopper next to the sluice feeds limestone aggregate into the drum, either manually or automatically, through a chute to the hub of the revolving drum. In self-feeding drum systems, a flexible drive shaft conveys power directly from the rotating drum shaft to the reciprocating feeder (Fig. 5-14). Small holes in the drum dispense the limestone into the stream.

Water volume through the sluiceway determines the speed at which the drums rotate and, consequently, the amount of aggregate supplied to the drum and ultimately to the stream. Thus, within the capacity of the drum, the amount of limestone needed to maintain a target pH/alkalinity value is available regardless of flow conditions. This dosage control principle is similar to that for a tipping bucket mechanism in a water-powered doser.

The limestone aggregate grinds within the drum in a wet autogenous process (i.e., the aggregate itself is the abrasive agent). The aggregate mixes with sluice water that enters the drums through tiny holes in the drum. Drum rotation causes abrasion, which in turn produces "fines" (90 percent of the particles are less than 30 μm in diameter) that are released into the stream through the same holes by which the water entered.

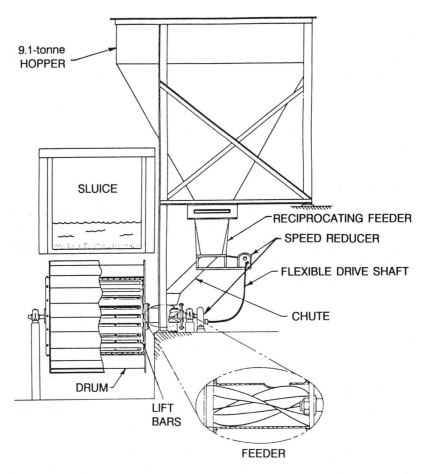

Figure 5-14.—Diagram of a self-feeding limestone rotary drum (Source: Olem, 1991).

Recent developments include screens inside the drums with various mesh sizes to control particle size. Output of the produced fines is controlled by aggregate size and rotational speed of the drum. In larger streams, several drums can operate in parallel.

The main design features of a self-feeding rotary drum system include a hopper, a feeding mechanism that varies with the speed of the drum, and a device that meters and transports limestone aggregates from the storage hopper into the interior of a water-powered rotary drum. A self-feeding rotary drum system was installed and put into operation in December 1988 on Dogway Fork in West Virginia.

Limestone aggregate can be fed continuously into the interior of the drum at the rate of 54 kg/h (1,200 lb/h). Therefore, at the optimal water flow of 0.113 m^3/sec (4 ft^3/sec) in the sluiceway, a single drum can produce a limestone dose of 15 g/sec (2 lb/min) or 133 g/m^3 ($8lb/1000ft^3$). A calibrated opening in the reciprocating feeder mechanism acts as a metering device to control the amount of aggregate fed to the drum during each stroke of the feeder. The higher the flow rate, the faster the drum turns. The number of strokes of the reciprocating feeder per unit time increases, and more limestone is supplied to the drum.

Bulk delivery of limestone can be taken annually and the limestone can be stockpiled at the site. The 10-tonne storage hopper for each self-feeding drum requires only weekly refilling from the reserves (usually with a front-end loader). Additional information is available in Zurbuch et al. (1991).

Costs of Stream Liming

Costs associated with the various liming techniques for streams described above are shown in Table 5-6. As before, the costs are converted to 1991 U.S. dollars. A major portion of the total cost for stream treatment is the initial purchase and installation cost of the treatment device. Additional costs include those of maintenance, labor, and the neutralizing material. Material and transportation costs for stream treatment are not included in Table 5-6 but can be estimated on the basis of costs shown in Table 5-2 and the dosage rate of the particular treatment device. A summary of the cost information presented for streams is shown in Table 5-6.

WATERSHED LIMING

If the lake cannot be effectively limed because of short water retention times, the application of limestone to the

Table 5-6.—Typical cost ranges for stream liming application techniques. [a]

APPLICATION TECHNIQUE	ANNUAL DOSAGE[c] (TONNES)	CONSTRUCTION COSTS			ANNUAL MAINTENANCE COSTS[b]		
		LOW	HIGH	AVG.	LOW	HIGH	AVG.
Doser: electrically powered	2,000	$44,000	$55,000	$50,000	$9,000	$13,000	$11,000
Doser: water powered	90	7,000	14,000	10,000	3,000	10,000	7,000
Diversion well	80	18,000	27,000	23,000	5,500	11,000	8,500
Limestone rotary drum	450	110,000	130,000	120,000	13,000	18,000	15,000

[a] Costs are based on actual liming applications after 1970 and are converted to 1991 U.S. dollars assuming 5% per year inflation.
[b] Does not include costs for purchase and transport of neutralizing material.
[c] To convert to tons, multiply by 1.103.

watershed should be explored. Unfortunately, only limited practical information on watershed liming is readily available; many reports are from Sweden and Norway and have not been translated into English. Watershed liming is recommended only if sufficient resources are available to review this information, develop a detailed plan, and carefully monitor the treatment activities and their effects. The reader is referred to reports from the Loch Fleet project in Scotland and summaries of Scandinavian research results. This information and appropriate references are described in Brocksen and Wisniewski (1990), Olem (1991), and Olem et al. (1991).

Timing of Liming

Liming strategies for watersheds may be influenced by whether the ground is frozen or if there is a significant snow-pack. If the ground freezes before snow develops and remains frozen during snowmelt, spring snowmelt may interact little with recently neutralized soils and thus preclude the possibility of reducing the acidity of the meltwaters. Also, it may simplify application particularly in northern climates to lime the snow-pack in winter. Limestone has been successfully applied to watersheds during each season. The choice depends largely on the particular site and its climate.

Location of Liming

Experience with watershed liming indicates the importance of applying the limestone principally to major water pathways. This practice reduces the area treated and the limestone required. Water pathways include wetlands, headwater streams, and, in general, areas of water discharge.

Discharge areas are the geographic portions of a watershed with surface flow, particularly during periods of stormflow or snowmelt. Limestone applied to these areas is dissolved by the water flow, which makes its way to the stream or lake that is the target for neutralizing. Liming recharge areas—dry soils— is either marginally successful or ineffective and considerably more expensive than liming water pathways. In dry soil liming, much of the limestone is tied up in the soils and does not dissolve in water that reaches the waterbody designated for neutralization.

Calculating the Dosage

No dose calculation procedures for watershed application are established, although models have been developed and tested in Sweden by Warfvinge and Sverdrup (see Olem, 1991). The limestone dose for watersheds varies based on factors similar to those that apply to lake and stream liming. The base saturation of soils is an additional factor important in determining the dose for watersheds. A general guideline is application of 5 to 10 tonnes/ha of limestone (2 to 4 tons/ac) to soils and 15 to 25 tonnes/ha (5 to 9 tons/ac) to discharge areas and wetlands.

Watershed Application Methods

Application techniques for liming soils and wetlands in the watersheds of streams and lakes are similar to methods used in agriculture and for liming lakes. Although actual application methods are not particularly difficult, how much and where materials are applied is critical.

Boat

Wetlands can be limed by using any of the boat application methods described earlier for lakes, provided the boat can be launched and operated in the wetland and the material can be distributed to where it is needed.

Truck or Tractor

Liming of forest soils or wetland areas by truck or tractor is also possible if the techniques described earlier for lakes are used. Various land vehicles can spread agricultural and pelletized limestone over portions of the watershed. Pelletized limestone may be useful to prevent adverse effects on vegetation and provide a timed-release of limestone to waterbodies.

Helicopters or Fixed-wing Aircraft

The techniques described earlier for lakes can be used to lime forest soils and wetland areas by aerial application. This method is usually more costly than using land vehicles but may be more feasible in areas where road access is limited. Aerial application may also distribute limestone to critical portions of the catchment—such as areas of groundwater discharge—that are difficult to reach for application by land.

Costs of Watershed Liming

Few sources of cost information for watershed liming are readily available. Based on limited sources of watershed and lake liming cost data, the cost for a single watershed liming operation is estimated at about five times that of direct lake application. If the lake retention time is 1 year and reacidification times for lake and watershed liming are 3 and 15 years, respectively, the economics of each technique are similar if considered on an annual basis. Because watershed liming may provide benefits not possible in most whole-lake liming efforts—such as treating feeder streams and protecting embayments from acidic episodes—the process may be warranted even if it requires more frequent treatments.

CHAPTER 6

Extra Help for the Habitat and the Fisheries

L iming alone may not be adequate to achieve some fishery management objectives. As Chapters 2 and 3 suggest, additional physical, chemical, or biological characteristics of the habitat can continue to limit fisheries after liming. In this chapter, we introduce a selection of methods to help identify some of the more common limitations and lessen the severity of many of them. The topics we discuss include combining fertilizer with lime to adjust water quality limitations, controlling or eliminating undesirable fish species that compete with the target species, stocking to enhance or maintain recovery of target fish species, and enhancing the physical habitat to improve recruitment and survival of fish.

ADDITIONAL WATER QUALITY MANAGEMENT OPTIONS

Waters that are highly sensitive to acidification not only have very low concentrations of dissolved chemicals that can buffer against acidification but also have very low concentrations of nutrients. These two factors generally limit biological productivity. Surveys of poorly buffered surface waters with sensitivities to acidification reveal that the waters with higher nutrient concentrations and higher productivity also have relatively higher resistance to acidification. This resistance occurs because the increased plant growth in these fertile waters leads to increased diurnal flux. The growth causes the release of dissolved organic carbon and higher diurnal concentrations of inorganic carbon compounds, which can buffer waters against changing acidities. Thus, for low alkalinity and low hardness surface waters, coupling fertilization to liming treatments can substantially reduce their longer-term sensitivities to acidification (Olem, 1991).

Fertilization experiments in oligotrophic lakes in Canada show that mild enrichment of nutrient concentrations can increase plant productivity without significantly altering the trophic status or structure of the biological community providing the food base for fish. Results from these experiments, for example, indicate that the community structures of free-floating algae (i.e., phytoplankton) in fertilized Canadian lakes are not altered greatly by the chemical treatment, while plant productivity is substantially increased (Stockner, 1981).

In general, the plant productivity of surface waters necessary for good fish growth is limited significantly when total inorganic nitrogen concentration is less than 0.4 mg/L and total phosphorus concentration is less than 0.01 mg/L. Lakes with nitrogen and phosphorus concentrations below these values are characterized as mesotrophic to oligotrophic.

When the goal of fishery management in an acidic water is to increase fish productivity and angling success and nutrient concentrations in the water are one-half to one-fourth or less of these levels, then fertilization can be considered as a useful supplement to liming. Fertilizing water to increase fish productivities should be considered only when concentrations of nitrogen and phosphorus do not exceed the concentrations cited. Higher concentrations may lead to nuisance blooms of algae and other undesirable water quality problems. When moderate fertilization is coupled with liming to manage the water quality of acidic lakes, fish productivity will increase and resistance to future acidification will be prolonged. The publications shown in Table 3-2 provide general guidance on the use of fertilization of small lakes and ponds.

CONTROLLING UNDESIRABLE FISH SPECIES

Surface waters in a fishery management and liming program may hold fish species that interfere with the established goal in managing for the target species. This can occur particularly when establishing new fish species in treated waters or when reclaiming waters currently populated with unwanted or undesirable fish species. These species may be undesirable, for example, in terms of low angler demand, offering unwanted competition with the target species for food resources, or potentially high predation rates on younger life stages of the target species. To advance more rapidly toward the management goal, the elimination of undesirable fish species *before stocking* limed waters may be useful. This can help maximize the growth and survival potentials for the target fish species.

For some smaller lakes and reservoirs, it may be possible to drain, siphon, or pump out the water and remove the undesirable fish. However, for most surface waters this is not possible, and the unwanted fish may be eliminated using a fish toxicant. The most common of these is rotenone, an inexpensive chemical that suffocates fish by blocking oxygen uptake at the gills. While it is relatively nontoxic to birds and most mammals (except swine) at treatment dilutions used for fish, care is required when using concentrated rotenone to avoid contacting the skin and, especially, the eyes, nose, and throat with the solution before dilution in the water.

Rotenone, a plant extract, is generally applied to surface waters as a wetted powder or liquid. Its toxic effectiveness can depend on the species and size of fish, water temperature, pH, and concentrations of dissolved oxygen and suspended matter. At water temperatures of about 50 to 75°F (10 to 25°C), 1.0 to 2.0 mg/L of rotenone is usually applied to assure a complete kill of all fish. Treatments in cooler streams and reservoirs tend toward the lower limit of this range (1.0 mg/L), while the upper limit (2.0 mg/L) is more common in larger, warmer, turbid lakes. This range of concentrations can also be lethal to zooplankton and many other aquatic invertebrates, although they will repopulate within a few weeks following treatment. Usually, affected fish begin surfacing within a few minutes to a few hours following application. Collection of these fish then can begin, continued for one to two days, or until the dead fish are no longer surfacing. Potential problems with fish toxicants can arise where high densities of weeds occur; these can trap treated fish and prevent their retrieval at the surface. Many of the publications shown in Table 3-2 discuss the use and availability of rotenone.

Rotenone often becomes nontoxic in the environment within a few days. It can, however, persist as toxic for over a month in deeper, cooler, dark layers of stratified lakes, especially in waters that are slightly acidic and have low dissolved oxygen concentrations (< 1 mg/L O_2). During the spring or fall overturn, these toxic concentrations of rotenone can be mixed into the upper water layers and cause toxicity again. Rotenone was applied as part of the fishery management and liming program in several lakes where extensive mortality of stocked fish occurred months after the rotenone and liming treatment. Therefore, when rotenone-treated lakes and reservoirs have deep, cold, slightly acidic water strata, waters from the deeper strata should be chemically analyzed to assure that latent rotenone toxicity will not pose a future threat to stocked fish.

Before application of the toxicant to limited parts of lakes, a treatment area, like the mouth of a cove, can be completely blocked off with netting, which will also capture the dead fish. In streams, the downstream bounds of the treated reach are also often blocked with netting. Sometimes, to neutralize the toxicant's effect, potassium permanganate can be applied outside the boundary of the blocking nets in standing waters and downstream of the blocking nets in streams.

The toxicity of rotenone can be reversed rapidly by adding 1.0 mg/L of potassium permanganate for each 0.05 mg/L of rotenone used. However, since potassium permanganate can also be toxic to some fish species at 3 to 4 mg/L, extreme care is needed to assure appropriate application rates. And, since this chemical is a hazardous substance, it requires the same care in handling that we noted previously for rotenone.

Before treating surface waters with any fish toxicants, it is vitally important to check with local authorities to assure compliance with applicable laws and regulations. In many states, it is illegal to apply or discharge any toxic substance to any public or private lake or stream without obtaining a permit from the appropriate state agency. Additionally, permission often must be obtained each time a private waterbody is treated with rotenone. In most states, detailed information on appropriate requirements and techniques for fish removal from private waters is available from local offices of state fishery management agencies.

STOCKING TO ACCELERATE RECOVERY OF FISHERIES

Fisheries management goals aim primarily at establishing self-sustaining fisheries, maintained through natural reproduction. Unfortunately, many lakes and streams lack appropriate spawning or rearing habitats and have harsh environmental conditions during spawning or other conditions that limit successful recruitment by the target species. Therefore, it is often necessary to augment or even to maintain the fishery totally by stocking. Stocking is especially important in those waters lacking current fish populations and is required when waters are renovated to introduce new fish species. Stocking also can be used to provide unique fishing experiences and to introduce new species.

Given the need to stock fish into many waters after liming, the following appropriate stocking procedures should be observed: obtain necessary permits to stock fish, establish appropriate stocking times and rates for the water, and identify sources for appropriate fish stocks. Additional details of local importance are available from state fishery management agencies about many topics we will discuss.

Legal Considerations

Before any fish are stocked, personnel at local offices of the state's fishery management agency should be consulted to obtain important information on local legal requirements. Private stocking of fish into private waters generally requires a permit from the applicable state fisheries management agency. Most states ban outright private stocking of public waters. In addition, many states require that any fish planted within its boundaries be certified free of disease.

In some areas, fish stocking is further restricted by special land use or other management plans. Common examples of this are wilderness areas where the lands are managed to preserve, enhance, or restore natural conditions. Additionally, stocking options may be limited for some waters by the presence of rare, threatened, or endangered species.

Alternative Fish Stocking Strategies

Where winter- or summer-kill problems are common or annual survival of fish is a problem, creel-sized fish must be stocked to maintain a put-and-take fishery. Larger fish can also be stocked to improve the quality of a catchable fishery. The higher costs for stocking larger fish can be offset by the higher harvest rates for these larger fish, especially at sites where angling pressures are relatively heavy. In fact, unless harvest rates are restricted in waters having heavy angling pressures, most stocked fish (up to 85 percent) can often be harvested within three weeks. When this occurs, the initially high satisfaction by anglers can rapidly turn to unhappiness and new demands for stocked fish.

If annual survival rates are not a problem but spawning habitat is limited or other factors interfere with recruitment of young fish to a population, stocking subcatchable fish in a put-grow-and-take fishery can be a cost-effective approach. Put-grow-and-take fisheries can also be a useful management strategy in some waters to supplement natural reproduction.

For waters having high angler pressures, stocking both subcatchable and creel-sized fish can be the optimal strategy.

This strategy supports greater angling intensities than plants using either size class alone. A Colorado study found that stocking fingerling rainbow trout did not alter catch rates for planted catchable trout nor did stocking creel-sized fish appear to affect size distributions of stocked fingerling fish (McAfee, 1984). Stocking both catchable and subcatchable rainbow trout did, however, increase the average year-to-year harvest of rainbows.

In limited situations, stocking new species to enhance the existing food chain in a managed water can be useful. Predator or prey species may be added to construct a more "balanced" system. Table 3-4 presents examples of appropriate combinations of fish species for stocking in managed surface waters.

Stocking Times

Selection of appropriate stocking times depends on the management strategy and species involved. For most put-and-take fisheries, stocking occurs in the late winter or early spring, before the period of angler use and when catchable fish stocks are available. For put-grow-and-take fisheries using spring spawning largemouth or smallmouth bass, most stocking occurs during the late spring or early summer since bass can be difficult to maintain for extended periods in culture facilities. Abundant weed growths or other protective cover in shallow waters can enhance survival rates for young bass by reducing predation rates.

More options are available for put-grow-and-take fisheries using either fall spawning brook trout or spring spawning rainbow trout. Either spring or fall stocking of fingerling fish of these species may be appropriate for limed surface waters since either group can reach catchable sizes at nearly the same time. However, in many waters, longer-term survival percentages of fall stocking can be greater as a result of lower metabolic demands and lower predation during the cooler season. For both trout species, the larger size of fall fingerlings also helps to increase post-stocking survival rates.

The relative cost of fish in the spring and fall is another consideration in the selection of stocking time. For example, costs of brook trout from commercial hatcheries in New York State in 1990 ranged from about $25 to $60 per hundred for spring fingerlings and $50 to $95 per hundred for fall fingerlings. The larger size of fingerling brook trout in the fall tends to decrease the number of fish stocked as well as increase their survival rates. Therefore, fall stocking tends to be more cost effective, since fewer fish are used.

State fishery management agencies can supply current information about where stocking fish are available. Most state-operated hatcheries do not supply fish for stocking privately owned lakes.

Stocking Rates

In most limed surface waters, maximum stocking rates and growth rates are limited because of their lower natural nutrient levels and shorter growing seasons. These factors indicate that stocking rates for most limed lakes should be significantly less than rates for lakes not sensitive to acidification within the same geographical area. For example, many lakes in the northern tier states can support fish standing crops reaching 225 to 450 kg/ha (200 to 400 lb/ac). But acid-sensitive lakes in the Adirondack Mountains of New York support maximum brook trout production levels of 7 to 16 kg/ha/yr (6 to 14 lb/ac/yr; Schofield et al. 1989). Therefore, brook trout in these lakes should be managed to maintain maximum standing crops in the spring of only 11 to 14 kg/ha (10 to 12 lb/ac). To achieve these goals, no more than 100 fall fingerling brook trout per hectare (40 per acre) should be stocked (Gloss et al. 1989a).

When excessive numbers of fish are stocked into lakes, high competition among the stocked individuals for the available food organisms can result in reduced growth rates, "stunting," and reduced survival percentages (Gloss et al. 1989a,b). Fish populations stocked in limed lakes that were previously fishless can rapidly decimate the accumulated populations of prey organisms. These organisms frequently make up a significant fraction of the total food resource available following stocking. When food becomes depleted, subsequent food limitations can produce important adverse effects on this survival and growth in limed lakes.

Studies of brook trout planted in limed lakes in New York's Adirondack Mountains clearly demonstrate these adverse effects (Gloss et al. 1989a,b). Mean individual growth rates for stocked brook trout were substantially depressed at the higher stocked densities and standing crops of brook trout (Fig. 6-1). At higher densities, the limited food resource was distributed among greater numbers of fish, restricting individual growth rates. Notably, the highest fish densities apparently depleted the food resource and caused the resident fish to lose weight. Lower stocking rates reduced pressure on the food resource, leading to significantly improved growth by stocked fish.

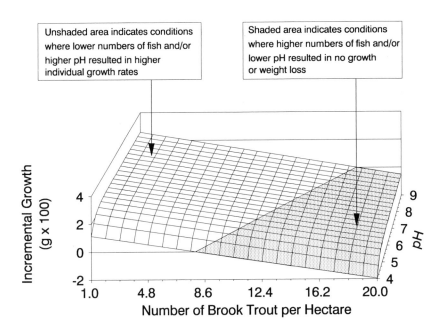

Figure 6-1.—Model predictions of incremental growth in weight (grams x 100) for individual brook trout related to field pH (calculated as H⁺ concentration in µeq/L) and estimated number of brook trout per hectare in Adirondack lakes (from Gloss et al. 1989a).

Restocking will also be necessary in many limed and fish-stocked waters. For example, most brook trout live only three to four years in many northeastern U.S. waters. Without natural recruitment of brook trout to limed lakes, they should restock at two-year intervals, when stocking of larger fall fingerlings is recommended for the reasons noted earlier. Stocking of relatively larger fall brook trout can lessen the loss of newly stocked trout through predation by those larger fish remaining from previous stockings. Maximum restocking rates for limed lakes in the mountain areas of the Northeast should be at 100 fall fingerlings per hectare every two years or 50 fall fingerlings per hectare every year (Gloss et al. 1989a). The latter approach can maintain a more "natural" mixture of fish in the lake as well as a more uniform year-to-year fishery available for angling.

Selection of Appropriate Fish Stocks

Selecting appropriate fish stocks for limed lakes includes evaluating the local availability of source stocks. As the distance between the hatchery providing the source stocks and the planting destination increases, so does the cost of fish. Increased travel time also increases potential transportation stress to the fish and potential mortality of fish both before and following stocking.

A second tier of considerations, which are equally important, is the potential sensitivity of the available fish to present and future expected qualities of limed waters. Because different fish species and different strains of the same species have different evolutionary histories, they often can have disparate susceptibilities to low hardness and acidic water qualities and dissimilar survival and growth rates after stocking. One recent study, for example, shows that splake, a cross between lake trout and brook trout, had greater survival rates in acidic Canadian lakes than did either of the parent species (Snucins, 1991). Other studies show that hatchery strains of brook trout, compared to wild brook trout strains, had poorer over-summer survival and growth rates in ponds in New York and streams in Wisconsin, although they had faster growth rates in hatcheries (Flick and Webster, 1964; Mason et al. 1967). Because of species and strain differences and because limed lakes frequently tend to reacidify, planting fish stocks thought to be less sensitive to potential impacts from acidity and avoiding those stocks thought to be more sensitive can be advantageous. The lower the stock's sensitivity to potential acidification, the longer its population will survive as the limed water begins to reacidify and/or as the population encounters episodic events of acidic runoff waters. Figure 3-1 presents the relative sensitivity to acidity of over 30 different fish species, based on field studies. Contact the local offices of state fishery management agencies concerning the availability of specific stocks and local strains of fish genetically adapted to acidic water qualities.

Fish stock selection for limed surface water should also consider angler desires. Selected stocks should be species that provide local interest to anglers. Also, different species and strains have dissimilar susceptibilities to angling, again because of different evolutionary histories. Various studies show the hatchery strains of many fish species and reproductive crosses of hatchery and wild strains can be much more susceptible to harvest than wild strains. And, brook trout are commonly viewed as a species highly susceptible to angler harvest. Because of these differences, stocks of fish used in managing limed surface water should match anglers' wishes to the extent possible.

Acclimation of Stocked Fish to the Stocking Water

Survival of some fish species stocked into limed waters may be increased by prestocking those acclimated to the water quality conditions. Acclimation of stocked fish to water temperatures in stocking waters is a longstanding practice. Acclimation can also be useful for increasing the initial survival rates of stocked fish where stocking waters have elevated acid and aluminum concentrations and where culture waters have markedly harder water qualities than the stocking waters (Flick et al. 1982).

Fish can be acclimated by exposing them to sublethal concentrations of acid and/or aluminum over several days before stocking. Laboratory studies show, for example, that acclimation of brook trout can significantly elevate their tolerances to the concentrations of acidity and aluminum that cause the physiological stresses found in unacclimated brook trout in waters contaminated with acid and aluminum (Wood et al. 1988a,b). Similarly, survival percentages of the fish cultured in hard waters can increase significantly if the fish are acclimated to softer waters before stocking them into softwater lakes, streams, and reservoirs. However, the long-term significance of chemical acclimation in reducing possible effects on fish following stocking is not entirely clear. Pre-stocking adaptation to acidic water conditions needs additional investigation before this strategy can be recommended for routine use in liming and fisheries management programs.

Additional Considerations for Stocking Fish in Limed Surface Waters

Two important considerations of toxic threat remain when stocking limed lakes and reservoirs. First, as we noted earlier in the section on controlling undesirable fish species, when a lake with a deeper, cooler, slightly acidic water layer is treated with rotenone during the liming-fishery management plan to renovate fisheries in the lake, the rotenone may persist in the deeper layer in potentially toxic concentrations, sometimes for months. Therefore, wherever rotenone has been used to treat lakes holding such water layers, it is important to analyze the water chemistry from the deeper layers to assure that the rotenone has sufficiently degraded before stocking fish.

The second potential toxic threat to stocked fish in lakes and reservoirs concerns the changes in metal solubilities that accompany liming. At high acidities (low pH levels), most metals are at their greatest solubility in water and their greatest potential toxicity. After liming, the solubility and potential toxicity of most metals decrease. However, in waters where metal concentrations are very high, toxicity can persist after liming.

Field observations of aluminum, in particular, indicate that changes in solubility following liming may temporarily increase its potential toxicity, as aluminum hydroxide can precipitate onto fish gills, causing death. Laboratory studies show that waters nearest the gill have relatively higher pH levels because of ion exchange activity at the gill. Under certain combinations of water qualities, the less acidic conditions at the gill can result in precipitation of aluminum onto gill tissue (Playe and Wood, 1990). This can occur as the relatively more acidic environmental waters containing higher aluminum concentrations are swept near the gills during respiration, causing the pH of this water to increase and aluminum to precipitate. Liming can produce such conditions when high concentrations of dissolved aluminum exist before liming. Therefore, when liming such waters, evaluating survival potentials in the limed waters for a few days before stocking can be valuable. A few individuals of the fish to be stocked can be held in a cage or holding pen in the treated water and their survival monitored. Chapter 7 discusses the use of such devices.

USING HARVEST REGULATIONS TO MANIPULATE FISHERIES

The most common approach to directly managing fish populations is regulating angling pressure by manipulating harvest regulations (Redmond, 1986). Five basic regulatory management approaches prevail: minimum fish size limits, limits on allowed angling times, maximum catch limits, slotted length limits, and catch and release. Fishing pressure on a given water often depends on its accessibility and visibility and its recreational facilities. Waters with higher angling pressures can require more extensive restrictive regulations to maintain good quality angling.

Typically, better angling quality is maintained by establishing minimum size limits, restricting the times of day or seasons when fishing is allowed, and setting maximum harvest limits. While the minimum size limits have been used historically to protect fish through the time of first spawning, fishery management now emphasizes protecting fish up to a balanced size, one large enough to maintain predation pressure on the forage base but not so large as to allow excessive natural mortality before the fish become available for angler harvest. For the most part, higher quality angling waters maintain a catch rate of greater than one fish per hour of angling effort.

Slot limits allow harvesting of fish in the intermediate size classes, which maintains better size distributions for the regulated population. When properly applied, this approach permits more fish in the population to reach larger sizes and is often used to establish trophy fishing waters. Under some slot limit regulations, one or a few larger fish may be harvested with a larger number of small to medium-size fish as defined by the "slot." Waters selected for establishing trophy fisheries should have growing seasons and productivity levels great enough to permit significant annual growth of the fish.

Where productivities are very low, any allowable harvest of larger fish will cause a significant reduction in the potential quality of angling. Therefore, catch-release regulations can enhance angling quality, as measured by catch rates per hour or by total catch of larger fish. Such regulations are particularly useful for waters where management goals place low emphasis on the fishery as a family recreational resource. Such regulations should not be applied, however, to waters with high rates of reacidification or annual periods with habitat conditions lethal to fish, such as conditions for summer- or winter-kill.

ASSESSING HABITAT CHARACTERISTICS LIMITING FISHERIES

A great diversity of habitat characteristics can affect fish community and population structures and productivities. Table 3-3 lists Habitat Suitability Index (HSI) models published by the U.S. Fish and Wildlife Service for 32 species of freshwater fish. These reports and additional information shown in Table 3-2 present the critical habitat requirements for each species. When the management plan for a limed water includes one or more of these species, the habitat associated with the treated water can be surveyed to determine whether the habitat meets the minimum requirement for the species before initiating any extensive management efforts. If critical requirements for habitat are lacking and use of alternate species is unacceptable, options exist for correcting some habitat limitations. We discuss some options at the end of this chapter.

In addition to the use of species specific information, assessing the total capacity of the habitat helps to establish the expected potential of a habitat for fisheries management. Several models can project fisheries productivity based on existing habitat characteristics or anticipated improvements. Two of the more commonly used models follow, one for lakes and reservoirs and one for streams.

Ryder (1965, 1982) found that fish productivity correlated highly with the ratio between the total dissolved solid concentrations in a lake and its mean depth. This ratio is termed the morphoedaphic index (MEI). Calculate the MEI by dividing the total dissolved solids, as measured in part per million (ppm = mg/L), by the maximum depth of the lake (in meters or feet). Then, to estimate annual fish harvest (H), use either equation 6-1 or 6-2:

$$H = 1.3825 \, MEI^{0.4458} \qquad \textbf{6-1}$$

where H = kg/ha/yr and
MEI = ppm/m, or

$$H = 2.094 \, MEI^{0.4461} \qquad \textbf{6-2}$$

where H = lb/ac/yr and
MEI = ppm/ft.

The relationship of the MEI to fish production has been investigated in lakes of diverse characteristics. The MEI often accounts for greater than 60 percent of the variation in fishery productivity (Jenkins, 1982). Applied to lakes included in a fishery management and liming program, estimates of fishery harvest based on the MEI provide rough estimates of the total potential yields possible from the treated water. This information can help define appropriate angler harvest restrictions for the treated water.

More than 100 models examined the relationship of fish standing crops to stream habitat characteristics (Fausch et al. 1988). One of these models, the Habitat Quality Index (HQI), is a multiple regression model that predicts trout standing crops from physical, chemical, and biological data (Binns and Eiserman, 1979):

$$\log(Y+1) = 1.12085 \, [-0.903 + 0.807 \log(X_1+1) + 0.877 \log(X_2+1) \quad \textbf{6-3}$$
$$+ \, 1.233 \log(X_3+1) + 0.631 \log(F+1) + 0.182 \log(S+1)]$$

where
- Y = predicted trout standing crop (kg/ha)
- X_1 = late summer stream flow (m^3/sec)
- X_2 = annual stream flow variation (m^3/sec)
- X_3 = maximum summer stream temperature ($^{\circ}$C)
- F = food index = $(X_3)(X_4)(X_9)(X_{10})$
- S = shelter = $(X_7)(X_8)(X_{11})$
- X_4 = nitrate nitrogen (mg/L)
- X_7 = cover (percent)
- X_8 = eroding stream banks (percent)
- X_9 = substrate (percent coverage by submerged aquatic vegetation)
- X_{10} = water velocity (m^3/sec)
- X_{11} = stream width (m)

While this procedure and similar models have been applied widely and with general success in diverse North American streams, there can be problems obtaining appropriate data to apply the model correctly (Scarnecchia and Bergersen, 1987). For example, the HQI includes qualitative judgments for variables X_7 (cover), X_8 (eroding stream banks), and X_9 (substrate). Inexperience in appropriately evaluating these variables can contribute to a poor performance of the HQI. Also, because the model was developed mainly using streams where the principal resident fish were rainbow, cutthroat, and brown trout, the model has limited applicability for most other fish species.

All existing habitat models have various problems. These include misapplication of statistical procedures, a lack of standard methods for measuring habitat variables, and limited potential for applicability. Consult the review by Faush et al. (1988) for problems concerning specific models. Most important, whenever a model is applied to guide management decisions, the assumptions and potential limitations of the model predictions must be thoroughly understood.

IMPROVING HABITAT COVER AND SPAWNING AREAS

Habitat structure provides cover for fish and their invertebrate prey and substrate for periphyton used as food by invertebrates. Increased structural diversity of the physical habitat in lakes and streams often directly correlates with increased densities and standing stocks of fish. Methods follow for enhancing the physical habitat of surface waters to improve fisheries.

Enhancing Physical Habitat Structure In Lakes and Reservoirs

The most common artificial structures in lakes use branches and trunks from small trees. These structures can be very durable, with some hardwood shelters lasting over 30 years in northern lakes. Structures suspended above the bottom are more durable than those placed on the bottom, and they are less affected by accumulations of settling silt. Brush shelters may be the most effective structures attracting fish (Fig. 6-2).

Rock reefs and automobile tires can also enhance lake basin structure, and both will last indefinitely. But, while tires are considered biologically nonpolluting, they are not aesthetically appealing (Fig. 6-3). Despite this drawback, some fishery biologists consider tire shelters very effective in enhancing biological productivity.

In reservoirs, submerged trees and brush can provide valuable feeding areas for fish and increase catch frequencies. Selective clearing of timber preserves and produces good fish habitat, but it can also produce obstacles for boaters, and dead tree branches are unattractive.

Artificial gravel beds can enhance spawning areas. The potential value of this option, however, is limited by sediment deposition and drawdown regimes in reservoirs. In reservoirs with fluctuating water levels, seasonal plantings of winter wheat on exposed shores, for example, can stabilize the shore and reduce erosion. When the plantings are reflooded, they can

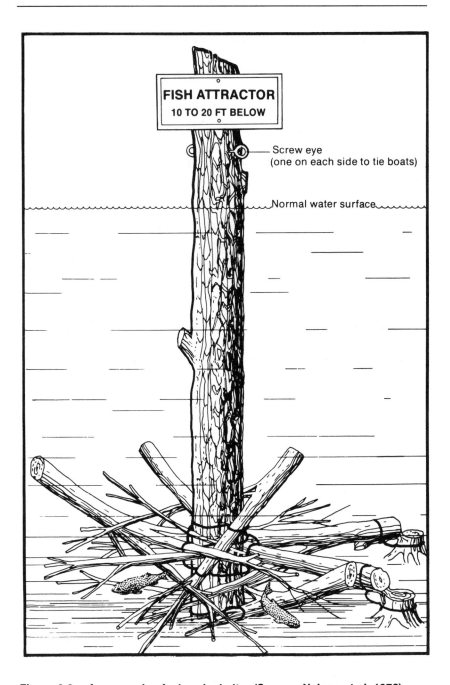

Figure 6-2.—An example of a brush shelter (Source: Nelson et al. 1978).

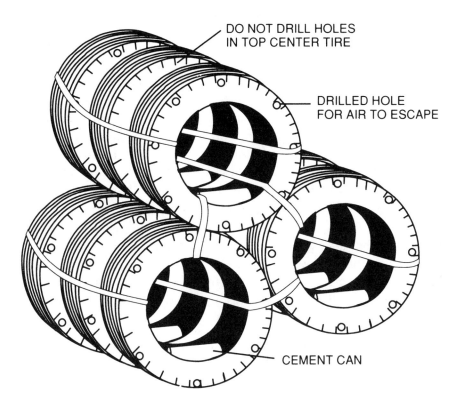

DO NOT DRILL HOLES
IN TOP CENTER TIRE

DRILLED HOLE
FOR AIR TO ESCAPE

CEMENT CAN

Figure 6-3.—One type of tire shelter (Source: Phillips, 1990).

provide spawning habitat, nursery areas, and nutrient releases that enhance nutrient recycling and productivity.

Most physical habitat enhancement efforts are directed at non-salmonid species in low-elevation, warmwater lakes. Despite this, brown trout, in particular, benefit from cover in streams, and brook trout also respond positively to cover improvements in mountain streams. Therefore, enhanced habitat structures in lakes and reservoirs might also benefit these and, possibly, numerous other species. Because many questions remain regarding the use of structures for any species, a structure placed in a lake or reservoir should be monitored to accurately evaluate its actual benefit.

Enhancing Physical Habitat Structure in Streams

Adding artificial structures is the primary approach for manipulating physical habitat in streams. Many of these structures are similar to those discussed for lakes. In use, these structures improve fisheries habitat two ways. They create new riffles and pools that (Fig. 6-4) reduce bank erosion and increase habitat diversity. For example, installing structures such as small rock dams and deflectors can increase the biomass of both fish and aquatic invertebrates and increase stream usage by terrestrial vertebrates.

The most commonly used within-channel structures are current deflectors, low profile overpour dams and weirs, bank cover, and boulder placement. Table 6-1 lists the common options for artificially modifying the physical structures of stream habitats. Nelson et al. (1978) and Wesche (1985) provide guidance on appropriate application, construction, and installation techniques for the structures.

Table 6-1.—Common artificial structures and management techniques used to enhance habitats for stream fisheries.[a]

CURRENT DEFLECTORS	
Rock-boulder deflectors	Half log deflectors
Gabion deflectors	Boulder placement
Double-wing deflectors	Trash catchers
Underpass deflectors	
LOW-PROFILE CHECK DAMS	
Rock-boulder dams	Gabion check dams
Single and multiple log dams	Beaver introduction
Plank or board dams	
BANK COVER TREATMENTS	
Log overhangs	Riprap
Artificial overhangs	Erosion-control matting
Tree retards	Stream bank fencing
Bank revegetation	Grazing control

[a] Compiled from Nelson et al. (1978) and Weshe (1985).

Fishery managers also use another approach for manipulating physical habitats in streams below reservoirs. Since the volume of water in streams ultimately defines the habitat available for fish, regulation of discharge waters from reservoirs can reduce physical limitations in downstream habitats. Holding back portions of the annual high spring flows and limiting discharge velocities during sensitive life stages of

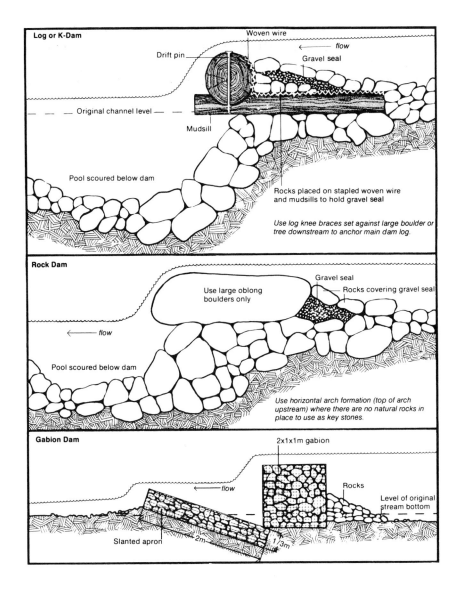

Figure 6-4.—Several types of check dams (Source: Nelson et al. 1978).

fish can reduce adverse effects accompanying high flows. If some of these stored waters are released to supplement low flows, adverse effects normally encountered during this period can be lessened. Also, the frequency and duration of high water (flushing) flows define the character of the stream bottom sediments and the availability of spawning habitat. Estes and Osborn (1986), EA Engineering Science and Technology, (1986), and Reiser et al. (1989) discuss a variety of methods for assessing instream flow needs for fisheries.

CHAPTER 7

Evaluating the Success of Liming

POST-LIMING MONITORING IS THE KEY TO LONG-TERM SUCCESS

Although follow-up monitoring to evaluate management actions is infrequent, it should be a required part of any environmental management effort, such as surface water liming. Evaluating the effectiveness of the management application and providing critical water quality information to show when additional management intervention is needed and reliming is necessary require monitoring information.

Practical management of limed lakes and streams focuses primarily on monitoring measures to determine whether treatment produced the desired changes in water quality and whether a healthy fishery is present. This chapter discusses simple monitoring measures that can easily satisfy evaluation needs by using these two criteria. For some treated waters and specialized requirements, especially where management goals are not met, resource managers may want more information on how specific waters or other biological groups are responding to treatment. For these needs, the following sections also introduce considerations and methods for expanding the monitoring program.

DEVELOPING PRACTICAL MONITORING PROGRAMS

Each surface water liming is, to a large extent, an experiment. Before treatment, the initial water quality conditions and volume of water to be treated are known as is the mass of limestone required to adequately reduce the acidity for an expected time. All liming efforts that follow are at the mercy of unknown future events. For example, unusually wet weather can rapidly flush treated water downstream, or unexpectedly high angling pressures can drive new populations of fish to extinction. Accurately assessing the success of management treatments requires monitoring.

Monitoring the success of practical fishery management and liming programs requires moderate data collection. In general, depending on the goal of the fishery management program, liming can be judged successful if its results meet any of three treatment criteria:

1. Development and maintenance of acceptable water quality conditions.
2. Establishment or maintenance of healthy fish populations.
3. Improvement of aquatic habitat conditions, including the quantity and quality of the forage base for the target fishery.

Evaluating the Chemical Habitat

In a practical monitoring program, appropriate measures are keyed primarily to the first two of these treatment criteria. As discussed in previous chapters, the two primary components of water quality targeted for change during liming are pH and alkalinity. Therefore, the same two chemical variables are of primary interest during monitoring. Is the pH 6.5 or greater and is the alkalinity 5 mg/L or greater? Results from water quality analyses completed before liming can be compared to other analyses completed after liming to assure that the application produced the predicted changes in water quality. As discussed in Chapter 8, times required for the dissolution of treated limestone can vary. Therefore, when liming lakes, measurements to record actual changes in pH and alkalinity should be completed, at a minimum, first about one week and then one month following liming. In streams, these measurements should be completed at increasing distances downstream, e.g., at 100 meters, 500 meters, 1.0 kilometer, 5 kilometers, and so forth.

The number and location of the sampling stations needed to evaluate the adequacy of treatment depend on the size and shape of the waterbody. When selecting sampling stations, one or two locations in the central portion of the lake or reservoir are often adequate. Also, where discrete coves exist, they should be sampled. At each sampling location, water from about 1 m (3 ft) below the surface and about 1 m above the bottom should be sampled to assess the effectiveness of vertical mixing of treatment material. Evaluating the effectiveness and duration of stream treatments can require three or more downstream sampling stations.

Over the longer term, chemical analyses in lakes and reservoirs should be completed, at minimum, in the spring and fall following seasonal mixing. A period of potential concern is mid- to late summer when stratification is strongest in the waterbody. Samples from the deeper water strata at these times can provide information on the development of water quality conditions potentially adverse to fish.

Between these periods of relatively extensive sampling across lakes, more frequent routine sampling can focus on water qualities of inlet and outlet streams. These analyses provide information on the general chemical condition of the waterbody by permitting comparison of water qualities in inflowing and outflowing waters.

Routine chemical analyses in streams should be conducted concurrently with the treatment application, usually during periods of rainfall and snowmelt when episodic runoff is occurring. These samples will help confirm that the treatment is effective through the reach of stream targeted by the management plan.

When fishery goals are not being met in a fishery management and liming program, and physical habitat characteristics appear to be satisfactory for the target fish population, other chemical variables should be evaluated. In particular, persistent toxic concentrations of heavy metals may pose problems for fish in some lakes after liming, as discussed in Chapter 8, or low concentrations of major or minor nutrients may limit production rates. When either of these conditions is suspected, chemical analyses in the monitoring program can be expanded to include additional chemicals of potential concern. Guidance for selecting and using appropriate analytical methods for these chemicals is provided later in this chapter.

Evaluating the Fisheries

Holding fish within enclosures in the treated water can confirm that water quality conditions following treatment are satisfactory for fish survival. As noted in the previous chapter, such practices should routinely be completed before stocking many limed waters to prevent extensive fish mortalities from latent quality changes.

Holding enclosures can be (1) wire mesh or fishnet mesh cages; (2) capped plastic six-inch sewer pipes with one-eighth to one-quarter inch diameter holes drilled abundantly along their lengths and into their ends; and (3) plastic garbage cans weighted with rocks to hold them on the bottom that have one-eighth to one-quarter inch holes in their sides to allow water passage. For smaller fish, plastic milk jugs with less than one-eighth inch diameter punched holes can be used (Fig. 7-1). Survival of the fish in these enclosures for three days to a week is generally adequate to indicate whether fish planted in the treated water will have a high potential for survival. To ascertain that appropriate procedures are followed during these on-site toxicity tests, additional enclosures with fish can be simultaneously placed in neighboring non-acidic lakes or streams as reference samples.

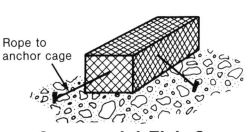

Commercial Fish Cage

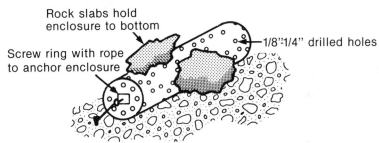

Capped Sewer Pipe Enclosure

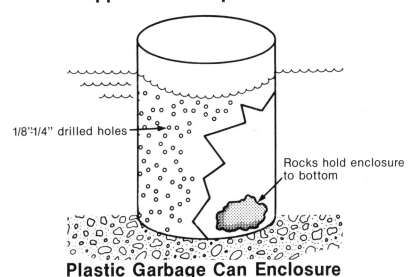

Plastic Garbage Can Enclosure

Figure 7-1.—Fish enclosures for use in monitoring.

The primary goal in most fishery management and liming programs is the establishment or maintenance of healthy fish populations and increased angler satisfaction. Assessing whether this goal is achieved is easily completed by using routine creel surveys. These surveys provide direct opportunities to assess the condition of individual fish in the waterbody, the year-class structure of the fishery, harvest rates for the population, and general angler satisfaction. Methods to conduct creel surveys are discussed later in this chapter.

When the fishery management goal is to establish a self-sustaining fishery, recruitment success is assessed by using minnow traps and minnow seines to sample shorelines or other water areas where young fish gather during the spring and early summer. These methods also are introduced later in this chapter.

COMMON ANALYTIC METHODS FOR PHYSICAL AND CHEMICAL VARIABLES

To assess whether a lake or stream provides acceptable conditions for fisheries requires data to characterize the physical and chemical habitat. Practical fisheries and liming management plans should include limited sampling to evaluate the habitat conditions present before and following treatment. The following discussion introduces commonly available methods to measure pH, alkalinity and acid neutralizing capacity, temperature, oxygen, color and dissolved organic carbon, transparency, major nutrients, metals, substrate composition and spawning substrate, stream flow and discharge, and lake volume.

Table 7-1 lists major suppliers for scientific equipment and supplies used for habitat analyses, and Table 7-2 presents expected cost ranges for representative equipment. Many chemical analyses most useful for assessing habitat conditions for fisheries are easily completed in the field by using specialized and relatively inexpensive test kits for water analyses. Such kits typically provide sufficient reactive chemicals to complete

Table 7-1.—Major sources for scientific equipment and supplies.

Fisher Scientific, 711 Forbes Ave., Pittsburgh, PA 15219; (412) 562-8300. General scientific laboratory and field equipment supplies.

Forestry Supplies, Inc., 205 West Ranking St., P.O. Box 8397, Jackson, MS 39284-8397; (800) 647-5368. Specializes in field monitoring equipment and supplies.

Hach Company, P.O. Box 389, Loveland, CO 80539; (800) 227-4224. Specializes in products for laboratory analyses and field test kits for chemical and physical water quality variables.

Hydrolab Corporation, P.O. Box 50116, Austin, TX 78763; (512) 255-8841. Specializes in multi-parameter water quality monitoring meters and recorders.

VWR Scientific, P.O. Box 5025, Sugarland, TX 77487; (800) 879-4100. General scientific laboratory and field equipment supplies.

Wildlife Supply Company, 301 Cass St., Saginaw, MI 48602; (517) 799-8100. Specializes in field sampling equipment and supplies.

YSI Incorporated, 1125 Brannum Ln., Yellow Springs, OH 45387; (513) 767-7241. Specializes in dissolved oxygen, pH, and temperature meters and probes.

Table 7-2.—Typical cost ranges for aquatic habitat and water quality sampling and monitoring equipment.

EQUIPMENT	COST[a]
Water sample bottle	$ 50 – 600
pH meter	50 – 500
pH test kit	25 – 50
Alkalinity test kit	25 – 150
Dissolved oxygen/temperature meter	250 – 1,250
Dissolved oxygen test kit	50 – 100
Color test kit	50 – 200
Secchi disk	25 – 125
Nitrogen or phosphorus test kits	20 – 300
Test kits for individual metals	25 – 300
Sieve sets for sediment separation	30 – 300
Water current/flow meters	200 – 3,000
Map measures	15 – 450
Planimeters	175 – 1,200

[a] Upper cost ranges for test kits generally include meters for colorimetric determination of concentrations. Other upper cost ranges are generally for research-grade equipment.

50 to 100 individual analyses, and each includes detailed directions.

Two potential problems exist in using these chemical test kits in the field. First, surface waters that are likely candidates for liming often have very low concentrations of most dissolved materials. The concentrations of some chemicals can be lower than these field methods can detect. Second, outside funding sources for some liming programs can require higher quality data than provided by many field test kits, including data screened according to EPA certified quality assurance/quality control procedures. At such times it may be necessary for a contract laboratory to analyze samples by using more exact methods. Representative costs for a selection of analyses by contract laboratories are shown in Table 7-3. Offices of state water quality regulatory agencies and regional EPA offices can help you find appropriate contract laboratories.

Table 7-3.—Representative costs for individual water quality analysis by contract laboratories (most costs decrease for multiple analyses).

PARAMETER	COST
Alkalinity	$10.00
Aluminum	12.00
Cadmium	10.00
Calcium	10.00
Copper	10.00
Lead	17.00
Mercury	35.00
Nickel	10.00
Nitrogen	
Ammonia	16.00
Nitrate	18.00
Organic	20.00
Total nitrogen	20.00
Organic carbon, total	30.00
pH	5.00
Phosphorus	
Ortho	15.00
Total phosphorus	17.00
Sulfate	12.00
Zinc	10.00

As discussed earlier in this chapter, most samples for chemical analyses should be collected at the sampling stations from waters 1 m (3 ft) below the surface and 1 m above the bottom. Water sampling devices appropriate for collecting water from discrete depths in the water column are available from suppliers listed in Table 7-1.

To collect water samples from discrete depths, these devices are lowered to the required depth by using a line with marked depth intervals. For most commercial water samplers, including those of the Kemmerer and Van Dorn type (Fig. 7-2), a weighted messenger is slid down the line to trigger the release on the end caps enclosing the water sample. Simple sample bottles easily constructed from readily available materials require only a quick jerk on the line to remove a stopper, which allows a weighted bottle to fill with water. However, the bubbling action from the enclosed air in this type of bottle makes the collected sample inappropriate for conducting dissolved gas analyses (e.g., dissolved oxygen) or for the highly precise pH and alkalinity determinations required for some projects. For other kinds of tests, the collected water can be poured into clean glass or plastic sample containers for analysis in the field or for return to the laboratory. Detailed information on sample collection and routine chemical analyses is available from U.S. Environmental Protection Agency (1983, 1987), Lind (1985), Standard Methods (1989), and Wetzel and Likens (1990).

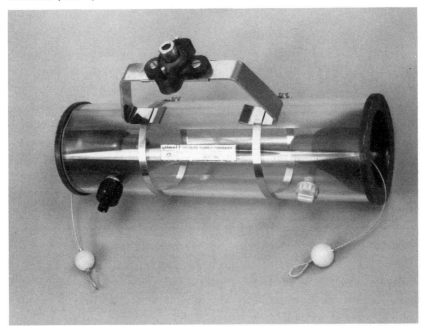

Figure 7-2.—Water samplers (courtesy of Wildlife Supply Co.)—(top) Alpha Bottle (Van Dorn sampler); (bottom) modified Kemmerer sampler.

pH

The pH of the sample water is usually determined in the field at the time of collection by using a pH meter and probe. The least expensive meters and pH field test kits are comparable in cost, accurate to 0.1 to 0.2 pH units, and adequate for most fishery management projects. When the highest precision is required for pH measurements, water samples are collected at the lake or stream stations in syringes and sealed. In the laboratory, syringe contents are injected into a closed chamber surrounding a highly sensitive pH electrode. When the sample is injected into the chamber, the pH meter connected to the electrode displays the pH of the sample.

Alkalinity and Acid Neutralizing Capacity

Alkalinity measurements of sufficient accuracy for most fishery management projects can be determined in the field by using test kits. The Gran analysis method, which determines the surface water's acid neutralizing capacity, is very precise. This technique measures not only the buffering capacity provided by dissolved carbonates, bicarbonates, and hydroxides (collectively included in alkalinity determinations) but also by organic acids and other compounds. Typically, acid neutralizing capacity is determined in the laboratory on samples collected and maintained in an amber high-density polyethylene bottle and stored at $4°C$ until analysis. These samples are titrated with a standardized acid, hydrochloric acid (HCl), and a standardized base, sodium hydroxide (NaOH). The pH measurements from the separated acid and base titrations are then analyzed using a modified Gran analysis technique to calculate acid neutralizing capacity (McQuaker et al. 1983; Kramer, 1984; U.S. Environ. Prot. Agency, 1987).

Temperature

Scientific surveys of waterbodies often measure water temperatures with electronic meters that are equipped with thermistors with long leads that allow temperature measurements in place at discrete individual depths below the surface.

Electronic meters must be calibrated according to the manufacturer's directions.

In practical liming and fishery management programs, however, such equipment may not be available. Instead, water temperatures at the surface can be measured simply by immersing a common thermometer into the surface waters. To determine temperatures at greater depths, sample bottles such as those shown in Figure 7-2 can be lowered to and filled with water from each appropriate depth. Then, the collected sample can be hauled rapidly to the surface and the temperature measured in the collected water. Knowledge about changes in temperature throughout the waterbody is needed to learn whether a lake or reservoir is stratified. Thermal stratification occurs when the middle portions of a water column include a layer in which the temperature decreases 1°C per 1 m increase in depth (about 2°F/3 ft). This layer is called the *thermocline* or *metalimnion*; the layer above it is termed the *epilimnion*, and the layer below is the *hypolimnion*. Stratification greatly impedes transport of dissolved and suspended materials among the three layers.

Dissolved Oxygen

Most healthy fish populations occur in waters with dissolved oxygen concentrations that exceed 3 mg/L. Field test kits are usually adequate to determine oxygen concentrations in the water. These test kits generally use simplified versions of the Winkler titration method (Lind, 1985; Standard Methods, 1989; Wetzel and Likens, 1990). In waters where thermal stratification occurs, managers must determine whether oxygen concentrations are significantly depleted in the hypolimnion by high rates of organic decomposition, particularly during the summer. If depletion is suspected, special care must be taken in collecting and analyzing deepwater samples to assure that they are not aerated in handling. Profiles of oxygen concentrations through the water column are often determined with an electronic meter equipped with a long-lead probe, which can be the same meter-probe combination used to measure temperature profiles. Again, when using such instruments, they should be carefully calibrated by following the manufacturer's directions.

Color and Dissolved Organic Carbon

Chapter 2 noted that naturally acid bog waters generally have an apparent color of greater than 75 platinum-cobalt units, caused by dissolved organic carbon concentrations of greater than 4.5 mg/L. This dissolved organic carbon includes tannins, humic acids, humates, and other organics that leach primarily from plant debris, including wood, needles, and leaves (Wetzel, 1983; Standard Methods, 1989). Color can easily be measured with field test kits. Some management programs may require direct measurement of dissolved organic carbon. Typically, samples for dissolved organic carbon analysis are collected, filtered, preserved to less than 4 pH with sulfuric acid, and stored at 4°C until analysis. Dissolved organic carbon analyses usually are completed at contract laboratories, which use a carbon analyzer calibrated according to manufacturer's specifications (Standard Methods, 1989).

Transparency

Transparency of surface waters is a relative index of lake productivity (Carlson, 1977) and an indicator of dissolution of limestone following treatment. In fishery management, transparency is commonly measured in the lake or reservoir by using a Secchi disk, a 20-cm disk that is either solid white or quartered into alternating black and white sectors. The Secchi disk is lowered into the water on a calibrated line until it disappears from view; then the depth is recorded. The disk is then raised until it reappears, and this depth recorded. The average of the two depth measurements is the Secchi disk transparency. The measurements should be taken in the shade of the boat to avoid glare on the water surface. Lind (1985) and Wetzel and Likens (1990) discuss methods for relating Secchi disk transparency to percent transparency and measurements of illuminated zone depths in water columns.

Major Nutrients—Nitrogen and Phosphorus

Nitrogen and phosphorus are the two nutrients that limit plant productivity in surface waters. Generally, increases in concentrations of either of these nutrients lead to increases in productivity. For comparison, surface waters having moderate to high productivities usually have concentrations of total inorganic nitrogen greater than 0.4 mg/L and total phosphorus greater than 0.01 mg/L (Wetzel, 1983). Results from field test kits usually provide adequate information on concentrations of these two nutrients for fishery management. For more exact results or to verify the results obtained from the field test kits, use a contract laboratory.

Metals

Field test kits are available for analyzing many metals (aluminum, calcium, chromium, copper, iron, lead, magnesium, manganese, molybdenum, nickel, silver, and zinc) and can provide valuable information that is useful in fishery management. We recommend, however, that analyses be conducted by contract laboratories when concerns exist about potential heavy metal concentrations in surface waters or fish tissues. Collection, preservation, and transport of the samples should be coordinated with the testing laboratory. Generally, water samples for total metal concentrations are acidified in the field to pH 2 by using nitric acid, whereas samples for dissolved metal concentrations are first filtered through a 0.45 µm membrane filter and then acidified to pH 2. Laboratory analysis is done with an atomic absorption spectroscopy (U.S. Environ. Prot. Agency, 1987; Standard Methods, 1989). An EPA (1989) report provides information for evaluating metal concentrations found in fish tissues.

Spawning Substrate and Substrate Composition

Good spawning habitat for many species often depends on the availability of substrate that contains particles of a size suitable for egg incubation. The best spawning areas for trout generally have gravel that ranges from about 3 to 75 mm (about 1/8 to 3 in). Optimal spawning areas for members of the sunfish family, including large- and smallmouth bass, bluegill, and crappie, usually include finer gravels and sands, most less than 3 mm (1/8 in) in size. The presence of silts and clays reduces the quality of spawning habitats for most species.

To evaluate the availability of suitable spawning habitat in most liming/fishery management programs, you must qualitatively assess the proportions of the area within the shallows of lakes and reservoirs or the total area of streambeds that are predominantly silts and clays, sands and fine gravels less than 3 mm in size, medium to large gravels 3 to 75 mm, and rubble and boulders greater than 75 mm. Adequate spawning habitat is available if greater than 5 percent of the substrate is available as sand and gravels for species of the sunfish family or as medium to large gravel for trout.

Stream Volume Flow

The volume of water flowing through streams (discharge) must be known to calculate appropriate dosing rates before adding limestone to streams. As discussed in Chapter 5, stream dosers are calibrated to the stages (relative depths) of the streams to deliver limestone at rates necessary for adequate neutralization. Volume flows are available for many streams from gauging stations operated by the U.S. Geological Survey or other groups. For these locations, flows are measured by reading stream stage (relative stream depth) from a staff gauge placed in the water. Staff gauges for smaller streams are often attached to the sides of rectangular-shaped metal flues that are set into the streambed so that the channel flows through a volume of known dimensions. Gauge readings are calibrated to relate stream stage to volume flow through a series of volume-flow measurements that occur at different stages.

Where calibrated staff gauges are not present, volume flows can be most exactly determined for individual cross-sectional areas in streams. Multiple cross sections can be measured and averaged to provide more exact measurements.

For each determination, the bank-to-bank width of the stream is measured, generally with a tape measure. Then the depth of the stream is measured at several intervals along the tape by a measuring rod (which can be a meter or yardstick in shallower streams) to provide a profile of the stream. Each depth measurement interval should not exceed 10 percent of the stream width (Fig. 7-3). At each of these intervals a current velocity meter is used to measure the flow rate of water at a point 0.6 of the distance from the surface to the bottom. The total volume flow or discharge is then calculated, based on the width and water velocity at each measurement interval, using equation 7-1:

$$Q = \sum_{i=1}^{n} (W_{i+1} - W_i) \left(\frac{d_i + d_{i+1}}{2} \right) \left(\frac{V_i + V_{i+1}}{2} \right) \qquad \text{7-1}$$

where Q = discharge or total flow (m^3/sec or ft^3/sec),
W_i = distance from left bank (m or ft),
d_i = water depth for each interval (m or ft), and
V_i = measured velocity for each section (m/sec or ft/sec).

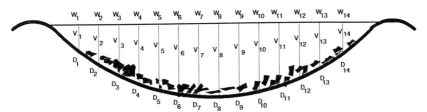

Figure 7-3.—Cross section diagram of a stream channel, showing measurements for widths (W_x), depths (D_x), and velocities (V_x) at 14 measurement intervals.

When a current velocity meter is not available, flow velocity can be more roughly estimated by using a neutrally buoyant object (such as an orange) to clock the time required for it to float through the middle portion of a measured 25 to 50 m (80 to 160 ft) of river. The time required for the float to travel the measured distance should be determined three times, using a stopwatch. The length of travel divided by the average of these times approximates the mean velocity for the reach (V_m). This technique is best where the reach of stream has a reasonably consistent cross-sectional profile, the dimensions of which can be determined by the same procedure. Total volume flows through the reach can then be roughly estimated using equation 7-2:

$$Q = \sum_{i=1}^{n} (w_{i+1} - w_i) \left(\frac{d_i + d_{i+1}}{2} \right) (V_m) \qquad \text{7-2}$$

where Q = estimated discharge or total flow (m^3/sec or ft^3/sec),
         w_i = distance from left bank (m or ft),
         d_i = water depth for each interval (m or ft), and
        V_m = mean velocity for the reach (m/sec or ft/sec).

When only rough approximations of stream volume flows are needed, as during a preliminary stream reconnaissance, for example, three depths across the stream at "representative" locations can be measured. The average of these values yields the mean depth. Also, the width of the stream is measured at a representative point along the reach. The product of the mean depth multiplied by the measured width provides an estimate of the cross-sectional area for the reach. This value can then be multiplied by the mean velocity of the reach (V_m), determined as described previously, to provide an approximation of the volume flow.

Lake Volume

Accurate dosing of limestone to lakes also requires a knowledge of the waterbody's volume. If the lake has a cone shape, its volume can be estimated using the formula for a cone's volume: half the distance across the lake is the radius (r) and its maximum depth is the height (h) of the cone (cone volume = $1.047r^2h$). If, instead, the lake is shaped like a sphere segment (i.e., a zone) and a radius of the segment surface equals s and the maximum depth equals h, then the formula for volume of a spherical segment can be used (segment volume = $0.5236h[h^2 + 3s^2]$). Standard surveying methods can be used to measure distances across lakes.

Because bottoms and shorelines for most lakes tend to be irregular, more accurate estimates of basin volumes can be based on the sum of the volumes contained in separate irregular conical segments (frustra) of the lakes (Lind, 1985;

Wetzel and Likens, 1990). These frustra can be pictured as stacked volumes of the lake bounded by each depth contour. If a map showing the depth contours is not available, contours can be determined by locating 10 or more transects across the lake, connecting readily visible landmarks. A boat is run along each transect and depths are measured at 10 or more equally spaced intervals along the transect with a sounding line that has distance intervals marked on a weighted line. A depth/fish finder also can be used to record depths along the transect. The depth measurements then can be plotted on a map showing the lake surface and transect locations and depth contours can be drawn on the map by extrapolation and connecting the points having equal depths; typically, for each 1 m or 5 ft depth interval. Using this plot, a planimeter can be used to measure the area of each depth contour. Estimated volumes for lakes are calculated by using equation 7-3:

$$V_t = \sum_{i-1}^{n} \left(\frac{h_i}{3}\right)(u_i + l_i + \sqrt{u_i + l_i})$$

7-3

where V_t = total estimated lake or reservoir basin volume (m^3 or ft^3),
  h_i = distance between the depth contours,
  u_i = area of frustra on upper contour (m^2 or ft^2), and
  l_i = area of frustra on lower contour (m^2 or ft^2).

COMMON METHODS TO ASSESS FISHERIES AND THEIR FORAGE BASE

Creel Surveys

Since an angling survey is the most common, simplest, and often the most effective and most enjoyable information collection technique for fisheries, creel surveys can provide useful information on fisheries and negate the need for other, more complex sampling strategies. Most simply, creel surveys require direct contact with anglers to record the length of time each angler fished and the total number of fish caught and their weights per angling hour for each species of interest during a defined sampling interval. While various approaches exist for conducting these surveys, the most easily applied creel survey techniques yield information for decisions on whether to lime and assess the success of fisheries management in lakes and streams.

These surveys can be stratified, i.e., split, to provide comparative information from weekday or weekend anglers. Three or more sampling intervals should be included in each period of interest; large sample sizes are needed for long periods of interest. Larger sample sizes also tend to reduce the error inherent in using samples to estimate population parameters. Therefore, the longer the period of interest, the larger the sample sizes required for accuracy.

Each creel survey sampling interval should be standard and should include as many anglers as possible who use the water during each interval. Also, intervals should be stratified to provide comparative results among morning, midday, and evening anglers. "Roving" contacts of anglers along shorelines or in boats and "point-of-access" contacts of returning anglers at boat launch areas or along access routes can be used alone or in combination. If all anglers using the water cannot be contacted directly during the survey interval, the total number of anglers active during the interval can be counted or otherwise estimated. Then, the information developed for sampled anglers can be extrapolated to all anglers fishing the water during the survey interval.

Minimum information from each angler surveyed should include time fished during the sample interval, number of fish caught (including those released) by species, and number of fish harvested by species. The survey can obtain additional useful information about the biological condition of the fish by measuring, weighing, and obtaining scale samples, when possible, from each fish harvested by the surveyed anglers. Survey information then is compiled and summarized to extrapolate total angler use (angler hours), total number of fish by species caught per angler hour, and total number and/or weight of fish by species harvested per angler hour. This information can then be compared to historical records for the water or to similar waters of interest in the region to learn whether angling success and harvest is, in fact, markedly below the norm expected.

When comparing among years and waterbodies, it is important to determine if compared data were developed using similar methods; they should also include similar sampling intervals.

Beach Seining

Beach or haul seines are useful in assessing fisheries during operational liming programs for lakes and some streams with slow currents. Seines are constructed of mesh panels; some include a bag in the middle portion to help collect captured fish. Floats generally support the top of the seine at the water surface, while a lead line holds the seine on the bottom. Beach or haul seines, which may range from a few feet to 30 m (100 ft) or more in length, have sticks or poles attached to each end of the seine ("two-stick seines") as handles.

Seines generally are fished by one or more individuals who wade as they hold each end. Generally, one end of a beach seine is first swept out from and then back to the shore in a wide, closing arc, while the opposite end remains stationary on or near the shore. Alternately, to increase the area of the arc sieved by the seine, the end of the seine nearest the shore can be moved parallel to the shore. To be most efficient, the area swept by the seine should have a depth less than that of the seine and a relatively smooth bottom, free of extensive weed cover. This allows the lead line to glide smoothly over the bottom, without riding up on obstructions that would permit fish to escape beneath the seine. Beach seines are most effective for capturing nearshore spawners and small, slow-moving, or schooling fish species that live in shallow waters.

In streams, when stream bottoms and banks are fairly regular, efficient seining can sample most species and size classes of fish. Seine from downstream to upstream when possible, preferably to the base of impassable falls or a strategically placed block net.

Field experience shows that long (30 m) seines cause less disturbance to small fish than do short seines and thus provide better estimates for the community of littoral-dwelling fish for the sample location. Seines of 6.4 mm (1/4 in) mesh limit the escape of smaller fish required for estimating recruitment success and the available forage base. The catch for a seine can be quantified in relationship to the bottom area included within the arc circumscribed by the seine haul.

Minnow Traps

Minnow traps, trap nets, fyke nets, hoop nets, and lobster, eel, and slat traps are examples of passive fishing gear. Minnow traps generally are baited with dried bread crumbs, cheese, or pressed soybean meal to attract fish. Wire or plastic minnow traps can be very useful in assessing the presence of small fish in lakes, reservoirs, and quiet areas of streams. Fish captured in these traps provide valuable information on the age, structure, and condition of smaller-sized fish species as well as information on the population structure for smaller individuals of larger fish species. Because of highly variable catch efficiencies, minnow traps do not provide very useful information on relative abundances of fish species. Since most of the fish captured can be returned to the water alive, passive fishing gear, in general, can serve as the primary means to capture fish in waterbodies where minimizing mortality is very important.

Other Biological Sampling Methods

The following chapter discusses generalized responses to liming found in most groups of aquatic taxa. However, for some limed waters, specific monitoring information may be desired about responses by various taxonomic groups or about aspects of fish populations in addition to those obtainable through methods discussed in detail in this volume. These kinds of information may be useful, for example, in evaluating why fishery management goals are not being achieved. Many useful references on these topics exist; Table 7-4 lists several commonly cited works and briefly notes the nature of the information in each reference. Consult these publications to obtain specific information about how to sample and assess information concerning specific taxonomic groups.

Table 7-4.—Useful references on methods for monitoring and assessing aquatic habitats and selected taxonomic references.

GENERAL

Standard Methods for the Examination of Water and Wastewater. 1989. 17th edition. American Public Health Association, Washington, DC. The classic standard source on methods for sampling and analyzing physical, chemical, and biological variables in all aquatic environments.

LAKES

Lind, O.T. 1985. Handbook of Common Methods in Limnology. Kendall/Hunt Publishing Company, Dubuque, IA. Text commonly used in many laboratory classes for training aquatic ecologists.

Weber, C.I., ed. 1973. Biological Field and Laboratory Methods for Measuring the Quality of Surface Waters and Effluents. EPA-670/4-73-001. Office of Research and Development, U.S. Environmental Protection Agency, Cincinnati, OH. Classic reference showing many EPA recommended methods.

Wetzel, R.G. and G.E. Likens. 1990. Limnological Analyses. 2nd ed., Springer-Verlag, New York. Text commonly used in many laboratory classes for training aquatic ecologists on methods used for lakes; includes limited discussion on streams.

STREAMS

Plafkin, J.L., M.T. Barbour, K.D. Porter, S.K. Gross, and R.M. Hughes. 1989. Rapid Bioassessment Protocols for Use in Streams and Rivers. EPA-444/4-89-001. Office of Water, U.S. Environmental Protection Agency, Washington, DC. Methods for assessing fish and benthic invertebrates in streams.

Platts, W.S., W.F. Megahan, and G.W. Minshall. 1983. Methods for Evaluating Stream, Riparian, and Biotic Conditions. GTR INT-138. Intermountain Forest and Range Experiment Station, Forest Service, U.S. Department of Agriculture, Ogden, UT. Compendum of recommended habitat and fishery assessment methods.

FISH

Everhart, W.H., A.W. Eipper, and W.D. Youngs. 1975. Principles of Fishery Science. Cornell University Press, Ithaca, NY. Classic text on methods used in fishery management.

Lagler, K.F. 1956. Freshwater Fishery Biology. W.C. Brown Company, Dubuque, IA. Classic text on methods used in fishery management.

Nielsen, L.A. and D.L. Johnson, eds. 1983. Fisheries Techniques. American Fisheries Society, Bethesda, MD. Recent compendium of methods used in fishery management.

INVERTEBRATES

Merritt, R.W. and K.W. Cummins. 1984. An Introduction to the Aquatic Insects of North America. 2nd edition. Kendall/Hunt Publishing Company, Dubuque, IA. Contains current methods for sampling aquatic insects and taxonomic keys to family and genus levels.

Pennak, R.W. 1989. Freshwater Invertebrates of the United States. 3rd edition. John Wiley and Sons, Inc., New York, NY. Recent edition of a class taxonomic work.

ALGAE AND LARGER AQUATIC PLANTS (MACROPHYTES)

Prescott, G.W. 1970. The Freshwater Algae. W.C. Brown Company, Dubuque, IA. Introduction to the collection and identification of algae.

Dennis, W.M. and B.G. Isom, eds. 1984. Ecological Assessment of Macrophyton: Collection, Use, and Meaning of Data. ASTM Special Technical Publication 843. American Society for Testing and Materials, Philadelphia, PA. Compendium of methods used for monitoring growth responses by larger aquatic plants.

Weitzel, R.L. 1979. Methods and Measurements of Periphyton Communities: A Review. ASTM Special Technical Publication 690. American Society for Testing and Materials, Philadelphia, PA. Compendium of methods for sampling and evaluating small plants. Very useful for monitoring effects.

THE NEED FOR RELIABLE DATA

This chapter concludes with a caution. As with most other tasks, successfully completing liming efforts and achieving fishery management goals depend on working with good information. The information is only as good as the data collected from the water before and after liming, both in large-scale scientific surveys and in practical fishery management and liming programs. Large scientific surveys typically have written procedures detailing all steps to be followed and double- or triple-check analytical results with duplicate analyses and, sometimes, splits of samples analyzed by outside experts or laboratories (Fraser et al. 1985; Brown and Goodyear, 1987; Porcella, 1989). While such detailed steps are not required for practical fishery management and liming programs, good records and accurate data are important. Managers must obtain the best data possible from all samples collected, analyses completed, and transects plotted and followed. If the program requires more than routine chemical analyses, extreme care should be taken to avoid cross-contamination of samples with waters from other sampling stations or depths.

Details about each sampling trip's procedures and problems should be recorded. When problems are encountered (samples dropped, nets lost, equipment broken, and so on), this information should be recorded. All records should be permanently maintained, which permits others to review exactly how the study was conducted.

CHAPTER 8

Physical, Chemical, and Biological Responses to Surface Water Liming

C hemical alterations produced in surface waters by liming lead to a cascade of interrelated physical, chemical, and biological changes in treated lakes and streams. These responses can vary over time and across different waters, even in some that appear overtly similar. The nature and magnitude of these changes define how liming will affect short- and long-term responses by fish populations. Therefore, when liming to benefit fisheries, you should have a basic understanding of the nature of the responses that can be expected, why they occur, and how they interrelate to help guide development of the best approach for managing water quality to enhance fisheries.

This chapter summarizes the physical, chemical, and biological changes that generally accompany surface water liming. In overviewing the general responses to liming, we have drawn from responses for a spectrum of treated waters. While the primary focus of this chapter is on lake limings, many of these same phenomena also occur in reservoirs. We also introduce several distinct responses for stream liming in a separate section. The following chapter discusses examples of managing acid surface waters with treatments applied directly into lakes and streams and on surrounding watersheds.

PHYSICAL AND CHEMICAL CHANGES IN TREATED WATERS

The initial change produced by treating surface waters with limestone is the physical increase in turbidity of the water. This increase in the number of solid particles suspended in the water column causes a short-term increase in shading and reduces light penetration into deeper water layers. Undissolved limestone particles also provide surfaces onto which organic carbon can adsorb from the water. When this is significant, the adsorbed DOC can markedly retard dissolution of limestone particles (Weatherley, 1988).

Initial chemical changes occurring with the dissolution of limestone include increases in pH (i.e., decreases in acidity), alkalinity, acid neutralizing capacity, and the concentration of dissolved calcium. Increasing concentrations for the first two of these measures are, in fact, the most often cited chemical goals for surface water limings.

As discussed more fully in Chapter 5, calcite, the mineral form of calcium carbonate, generally should comprise at least 70 percent of the chemical composition of limestone. Calcite supplies most of the chemical neutralization provided by dissolution of limestone. Strong acids normally dissociate to their ionic constituents in water. For example, sulfuric acid (H_2SO_4) dissociates into hydrogen ions (H^+) and sulfate ions (SO_4^{2-}):

$$H_2SO_4 \rightarrow 2 H+ + SO_4^{2-}. \qquad \text{8-1}$$

The free hydrogen ions cause water to become acidic. Following liming, calcite reacts with such acids, water molecules form, and carbon dioxide can be released to the atmosphere:

$$CaCO_3 + 2\,H^+ + SO_4^{2-} \rightarrow CaSO_4 + H_2O + CO_2. \qquad \textbf{8-2}$$

Generally during liming, however, most of this CO_2 does not enter the atmosphere. Instead, most of it dissolves into the water, reacting to form carbonic acid (H_2CO_3). In fact, atmospheric CO_2 in equilibrium with distilled water at sea level causes the acidity of water to stabilize with a hydrogen ion content near pH 5.6. At pH levels of greater than 4.5, some of this carbonic acid dissociates, releasing hydrogen and bicarbonate ions to the solution:

$$CO_2 + H_2O \rightarrow H_2CO_3 \rightarrow H^+ + HCO_3^-. \qquad \textbf{8-3}$$

It is the hydrogen ion from this dissociation of carbonic acid, as well as from the dissociation of strong acids, that provides most of the acidity necessary during the dissolution and reaction of the calcite contained in limestone particles:

$$H^+ + HCO_3^- + CaCO_3 \rightarrow Ca^{2+} + 2\,HCO_3^-. \qquad \textbf{8-4}$$

To summarize to this point, the chemical reactions shown in both Equations 8-2 and 8-4 cause the neutralization of acidity by removing free H^+ from the system. The reaction shown in Equation 8-4 also leads to an increase in bicarbonate and, consequently, an increase in alkalinity in the system. In Equation 8-3, the increase in acidity accompanying the release of H^+ and the increase in alkalinity accompanying the release of HCO_3^- balance each other.

Dissolution of calcite during liming (represented by the combination of Eqs. 8-3 and 8-4) creates a short-term high demand for CO_2. This demand can deplete free CO_2 from the water and delay complete dissolution of the added limestone until CO_2 dissolved in the water is resupplied from the atmosphere or from deeper water layers. Depending on various factors, including limestone dose rate, its particle size, water temperature, lake stratification, and other considerations more fully discussed in Chapter 5, complete dissolution may occur over a few days or take as long as several months.

The pH increase in limed waters also causes a shift in the predominant species of dissolved inorganic carbon. As suggested by Equation 8-3, progressive increases in acidity levels above about pH 4.5 causes the dissolved inorganic carbon to shift increasingly to greater proportions of HCO_3^-. For example, at pH 6.4, about equal concentrations of free CO_2 and HCO_3^- occur in water; whereas, between pH 6.4 and 10.33, HCO_3^- predominates in the dissolved inorganic carbon (Fig. 8-1). At higher pH levels, which are not commonly encountered in most natural surface waters, hydroxide ions (OH^-) increasingly predominate.

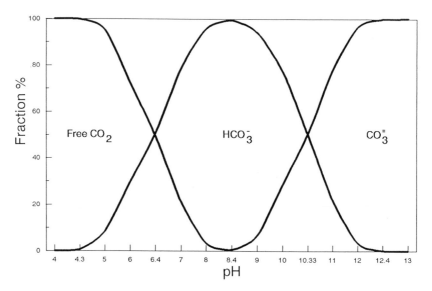

Figure 8-1.—Relation between pH and the relative proportions of dissolved inorganic carbon species.

The increase in pH also shifts the chemical species and solubilities for other chemicals in limed waters (Driscoll et al. 1989). Perhaps the most important of these is aluminum. Concentrations of the soluble inorganic monomeric aluminum (primarily, Al^{3+} and $Al(OH)_x$) decrease to negligible concentrations as the pH of water is increased to 6.0. The inorganic monomeric form of aluminum is the most toxic to aquatic organisms. Thus, concerns about aluminum as a significant source of potential toxicity in surface waters decrease as pH increases to 6.0 and above. Similarly, pH increases also generally decrease the solubilities of many toxic heavy metals, while increasing portions of less soluble species. Hence, total concentrations of most metals decrease in limed waters as the less soluble forms of these metals settle from the water to accumulate in the sediments.

Removal of aluminum from the water often increases the availability of phosphorus to plants, as proportions of phosphorus removed by precipitation of relatively insoluble aluminum-phosphorus compounds significantly decrease (Jansson et al. 1986). In addition, cycling rates for many nutrients generally show long-term increases in limed surface waters. As discussed in the following sections, this appears to be the result of more favorable acidity levels that allow increases in microbial and bottom-dwelling (benthic) invertebrate populations and leads to accelerated degradation rates for organic materials. Overall, many of the chemical changes produced by liming improve conditions for fish life (Marcus, 1988).

RESPONSES BY BACTERIAL POPULATIONS

Bacterial populations in both the water column and the sediments generally have marked long-term population increases after dissolution of the limestone and an initial period of short-term decrease in population densities (to be described). These increases are largely attributable to post-treatment pH levels that are generally favorable to many bacterial species. This favorable environment often leads to intermediate or long-term increases in decomposition efficiencies, nutrient cycling rates, and overall productivity levels in the treated system, including higher rates of fish production.

Part of the large initial increase in nutrient availability and general increase in system productivity following liming results from accelerated decomposition rates for the excess detritus, which often accumulates in acidic waters (Weatherley, 1988). That is, decomposition rates decrease in acidified environments and accumulation rates for detritus in the sediments increase. After liming and increases in environmental pH, the accompanying increases in bacterial numbers and species richness contribute to accelerated decomposition and disappearance of the excess accumulated detritus. Over the long term, after a store of detritus is reduced, nutrient cycling and productivities often decline to lower normal rates.

Occasionally, liming can cause rapid, short-term decreases in bacterial populations (Weatherley, 1988) for two primary reasons. First, metal precipitates produced by increasing pH can adsorb bacterial cells, removing them along with the metal precipitate particles settling from the water column. Second, calcite particles in the water column can adsorb dissolved organic carbon, which results in calcite particles competing with bacteria for organic carbon, thus reducing the food resource available to bacteria. Additionally, some acidic waters can have high concentrations of metals before liming (e.g., lakes affected by smelter emissions). Metals that precipitate from these waters after liming can produce high concentrations in sediments. Subsequently, these accumulations can be toxic to bacteria, causing long-term reduction of the bacterial population and slow nutrient cycling rates in the sediments. Potentially, such short-term bacteria reduction can lead to a lower productivity rate for fish. However, this occurs infrequently.

RESPONSES BY PHYTOPLANKTON

Population densities of microscopic, free-floating algae (phytoplankton) in acidic lakes are often markedly reduced immediately following liming (Bukaveckas, 1989; Molot et al. 1990). The cause of these short-term reductions, which may last about a month, is not totally understood, but it may be affected by the depletion of CO_2 from the water column during dissolution of calcite. Also, it may result partly from increased shading of the water column by the limestone particles that reduce light penetration to the deeper water layers. Both CO_2 and light are essential for photosynthesis, and significant reductions in either can lead to decreased plant growth or death. Additionally, initial reductions in phytoplankton standing crops from the upper water layers may result in part from loss of algal cells that co-precipitate with settling metals and to increased grazing by zooplankton.

Algal populations that were resident in pre-treatment waters can be stressed and decline as the pH increase causes the predominant chemical species of dissolved inogranic carbon to shift from free CO_2 to HCO_3^-. While both free CO_2 and HCO_3^- can be carbon sources for algae in photosynthesis, not all algal species are equal in their ability to use each of these inorganic carbon sources. Algal species that are obligate users of free CO_2 will be at a physiological disadvantage and excluded from neutralized environments where HCO_3^- predominates. Further, those algal species that can use both carbon sources require time to shift internal physiological mechanisms from free CO_2 to HCO_3^--based metabolism. Rapid increases in pH can occur over a period too short to allow successful completion of this shift in some species. Therefore, algal species with slower rates of metabolic adaptation to HCO_3^- can be lost from the system following limings, at least temporarily.

Over the long term, primary productivity and algal species richness generally increase following liming (Weatherley, 1988). This increase appears related primarily to the chemical changes produced by liming. They create more favorable water quality conditions for phytoplankton and increases in nutrient cycling rates; they also reduced competition with dissolved aluminum for phosphorus in the post-treatment waters. However, the composition of phytoplankton populations in limed lakes generally displays greater year-to-year variability compared to populations commonly found in neighboring, naturally circumneutral (near pH 7) lakes. This difference results largely from an increasing number of new algal species that are better able to colonize treated lakes following reduction of acid stress. The continued colonization causes continual adjustments in competitive forces among the resident species, with accompanying variation in population sizes. Additionally, grazing pressure on phytoplankton also can vary during this period, as stocks of zooplankton and fish adjust to the post-treatment environment. Not enough long-term information exists on whether the phytoplankton in treated lakes will eventually achieve population patterns similar to those found in neighboring circumneutral lakes.

RESPONSES BY ZOOPLANKTON

Zooplankton are microscopic animals, most with only feeble swimming abilities. As with phytoplankton, zooplankton populations often decrease markedly in the initial period following liming (Weatherley, 1988; Brett, 1989). The initial, short-term losses of these animals undoubtedly result directly from pH shock from rapid reduction in post-treatment acidity, and the short-term loss of their bacterial and phytoplanktonic food sources from the water immediately following treatment. Zooplankton communities often show significant recovery within a week of treatment, and some species have shown tremendous population increases within a month. Often, zooplankton abundances return to or exceed pretreatment levels within a year. While community compositions of the zooplankton species can differ between the pre- and post-treatment communities, this is not always the case.

The extent and duration of the initial effects on zooplankton from liming also appear to be affected by the season during which treatment occurs. The effects of fall treatments tend to persist longer than those from spring treatments. This appears to be related to the typical seasonal patterns for zooplankton biomass, which increases in the spring as phytoplankton biomass increases and decreases in the fall as phytoplankton decrease.

Other complicating factors that can affect zooplankton populations in treated waters include the presence of other toxicants (e.g., heavy metals found in some lakes affected by smelter emissions and mine drainages) that can continue to depress zooplankton numbers. In addition, where lake fertilization accompanies liming, massive increases in biomasses of zooplankton can occur. Further, post-treatment changes can include increased and altered predatory habits on zooplankton by dynamically changing fish populations.

Stocking fish after neutralizing previously fishless lakes can dramatically affect zooplankton communities (Weatherley, 1988). The dominant planktivore in fishless, acidic lakes is most often a larger invertebrate, which is very susceptible itself to fish predation. Invertebrate predators generally consume smaller zooplankton forms, causing larger forms to dominate in the zooplankton community. When fish that prey on zooplankton are introduced to these lakes, the larger forms incur heavy predation pressure; consequently, smaller forms of zooplankton tend to become dominant. This can cause the diversity of zooplankton species to increase in some treated lakes because reduced densities of larger zooplankton species lower competitive pressure, allowing increasing numbers of smaller zooplankton species to expand their population.

As with phytoplankton communities, zooplankton communities in treated lakes can require very long periods to develop equilibrium communities because new zooplankton species are continually able to colonize as a result of the improved water qualities. In summary, zooplankton show very mixed responses to liming. These complicating factors often make it difficult to project or interpret their responses accurately.

RESPONSES BY LARGER AQUATIC PLANTS

Larger aquatic plant species (macrophytes) provide important sources of habitat cover and food for fish and many populations of aquatic invertebrates. Neutralization of acidic surface waters appears to have minimal short-term effects on most of these species. However, a few plant species that favor low pH environments can show long-term population decreases in response to liming (Weatherley, 1988; Roberts and Boylen, 1989; and Jackson et al. 1990). These include some species of rushes (*Juncus* spp.), most *Sphagnum* moss species, some mat-forming benthic blue-green algae, and a few filamentous green algae such as *Mougeotia*. *Sphagnum* mosses, for example, are generally unable to use HCO_3^- in the water as a carbon source for photosynthesis. Therefore, when the predominate chemical species of dissolved inorganic carbon shifts from free CO_2 to HCO_3^- as pH increases following treatment, *Sphagnum* growths can disappear from shorelines within one to two years.

Over the longer term, some species of larger aquatic plants can increase in abundance and productivity because of the favorable decrease in environmental acidity and increase in nutrient cycling rates. Colonization by new plant species and changes in phytoplankton community densities can alter the nature and extent of interspecies interactions among the various plants. For example, where phytoplankton densities increase sufficiently to shade submerged macrophytes, loss of the larger plants can result. While the long-term consequences of liming on the community of larger plants appear highly variable among the treated lakes, no adverse long-term effects have been reported.

RESPONSES BY BENTHIC INVERTEBRATES

Benthic invertebrates, particularly aquatic insects, provide important food sources for fish. Liming can cause short-term decreases in some benthic invertebrate species and no effect on others. When liming produces high concentrations of toxic metals in the sediments, it can depress benthic invertebrate populations. In most lakes, however, where complications from toxic materials in the sediments do not occur, improved chemical conditions and increased food production rates following liming can lead to long-term increases in benthic invertebrate populations and species richness (Nyberg et al. 1986; Weatherly, 1988; Evans, 1989; and Keller et al. 1990).

New benthic species naturally colonize treated waters most often by migrating from nearby lakes or streams. But natural sources of invertebrates may not be equally available for all lakes. One study found that new species had not colonized treated lakes during two years after liming. In these lakes, new benthic species were artificially introduced and eventually established. The success of invertebrate stocking programs is not equal for all species. In neutralized lakes generally, increases in the numbers and kinds of benthic invertebrates produce greater grazing, shredding, and mixing activities and increase nutrient cycling rates.

RESPONSES BY FISH

Of the potential environmental effects caused by acidification, adverse effects on fish rank among the greatest public concerns. Consequently, considerable study has focused on these effects. Chapter 2 introduced the effects of surface water acidification on fish, and discussion follows on the benefits of liming for fish populations.

Surface water liming can directly benefit fish by reducing the water acidity, decreasing its toxic aluminum concentration, and increasing its calcium concentration. Reducing these sources of chemical stress on fish can markedly increase the potential for successful reproduction and recruitment of new fish into populations (e.g., Schofield et al. 1989; Gloss et al. 1989a, b; Gunn et al. 1990). Of course, other chemical stressors or physical limitations of the habitat, such as limited spawning areas discussed in Chapters 2 and 6, can continue to limit potential increase.

Following liming, growth rates of individual fish often increase for several reasons. First, the release from chemical stress allows fish, which may have had their feeding behavior and metabolic activity suppressed by acidic waters, to return to normal feeding patterns. Concurrently, the release from stressful chemical conditions can also lead to increased productivity by other organisms, thereby increasing food resources for fish. Fish introduced in formerly fishless waters with abundant food supplies can show phenomenal initial growth rates.

Following initial short-term bursts of growth, fish growth can slow considerably over the longer term as abundant food resources are consumed. This occurs because surface waters sensitive to acidification not only have naturally low acid neutralization capacities (i.e., alkalinities), but they also have naturally low nutrient concentrations and, consequently, low biological productivities. Therefore, when lakes are stocked with fish following liming, stocking densities must be balanced against the carrying capacities of the lakes (see Chapters 3 and 6). Excessive stocking densities can produce high competition among the resident fish that result in low growth rates (i.e., stunting), reduced survival, and undesirable fisheries in the treated lakes.

In some limed lakes, particularly those affected by smelter emissions or mine drainages, potentially toxic metal concentration can persist for months following liming. When fish are stocked into such lakes before these metals cease to be sources of toxicity, the rapid death of the stocked fish can result. Therefore, to help assure fish stocking success where problems from metal toxicity are likely, monitoring dissolved concentrations of metals in the water or directly evaluating the toxicity of the water by in-site toxicity testing methods is important.

Fish in some lakes have also been found to accumulate mercury in their flesh after liming. Here again, where such problems are expected, it is useful to monitor increases in mercury in the fish tissue. Additional considerations useful for managing fisheries in limed lakes, including acclimation, adaptation, and other genetic considerations, are discussed in Chapter 3.

RESPONSES ACCOMPANYING STREAM LIMING

Many of the same physical, chemical, and biological benefits that accompany lake liming can also accompany stream liming (Ivahnenko et al. 1988; Olem, 1991). The principal differences occur because lake limings can often result in residual treatment effects over extended times, whereas stream limings require strategies that provide continuous treatments during episodes when runoff of acidic water enters streams from their surrounding watersheds. Between such episodes, treatment of many streams may not be necessary.

This on-again, off-again approach required for stream limings can lead to a variety of problems and responses, especially when liming is done improperly or automatic treatment equipment malfunctions. For example, if insufficient neutralizing material is dosed into a stream during an episodic runoff of acidic waters, acutely toxic chemical conditions may persist that could lead to the death of resident organisms. This can set back any previously successful efforts to reclaim the stream. Such problems may be particularly prevalent during periods of extreme high flows (e.g., 200-yr flood events) that can exceed the capabilities of the treatment system design.

In comparison, a treatment system malfunction that doses excess neutralizing material to streams can produce chemical stress if the material used is not limestone and permits high pH values to be reached. It can also cause physical stress if the dosed material layers over stream bottom sediments. Coated sediments can cause deaths of insects, other organisms, and incubating fish eggs resident in benthic habitats.

Despite these potential difficulties, stream liming remains an important method for protecting stream habitats and their biological communities from acidification. As with lakes, such protection can be important for habitats containing sensitive, economically important, and culturally important species, including those that may be viewed as threatened or endangered. As discussed in Chapter 5, several approaches exist for managing water qualities in potentially acid-sensitive streams.

GENERAL CONSIDERATIONS

A few additional important points remain about general responses to neutralizing treatments for acidified surface waters. First, while various treatment methods and chemicals are available for neutralizing surface waters, increasing the pH ($\geq$ 6.5) and reducing toxic concentrations of aluminum and other toxic metals are only two of the primary goals for such treatments. The third important goal is to increase dissolved calcium concentrations. As discussed in Chapter 3, calcium concentrations less than 4 mg/L can be very stressful to many fish and other aquatic species, even without other sources of chemical stress. Also, increased ambient calcium concentrations help to protect aquatic organisms from physiological stress that occurs when acid and aluminum concentrations increase as lakes and streams reacidify.

This leads to a second important point. Surface-water liming is frequently only an interim solution. Often, acid-forming materials will persist for years within watersheds surrounding the more recently acidified lakes and streams. Therefore, waters draining from these watersheds will continue to reacidify surface waters. Such reacidification processes will largely reverse the responses discussed in the previous sections. Therefore, the decision to begin liming a lake or stream must include the decision to continue treatments into the future.

The give and take between neutralization and reacidification responses in treated surface water often produce unstable habitats for resident organisms. Of course, non-acid surface waters do not always provide stable environments for aquatic organisms. Seasonal changes and extreme weather events can produce very large changes in all aquatic environments. But, compared to environmental conditions found in neighboring surface waters where acidification forces are not a prevailing factor, the extra chemical variability created by repeated cycles of acidification and neutralization leads to an additional set of forces that affect the biological communities in neutralized lakes and streams.

Many acid-sensitive life stages and organisms can inhabit post-treatment waters, and many bacteria, plant, and invertebrate species often colonize these waters freely. However, the tendency of treated waters to experience brief periods of acid stress and to reacidify between neutralization treatments continues to stress acid-sensitive species and life stages. This stress can cause loss of colonizing species from lakes and streams.

Treated surface waters, therefore, not only can have unstable acid-base chemistries, they can also have continual changes in the composition of their biological communities. Consequently, treatment protocols to neutralize acidic lakes and streams rarely lead to ecosystems that are the same as those existing before initial acidification. Instead, treated lakes and streams will, at best, produce ecosystems that are structurally and functionally similar, but not identical, to the original systems. The changing acid-base chemistries, colonization by new species, recurring loss of acid-sensitive species, and continually changing interactions between resident species will produce "non-equilibrium" systems.

Finally, not all acidic aquatic habitats need to be neutralized. Naturally acidic ecosystems have long been part of our environment. Softwater lakes, acid bogs, and streams draining these environments provide important habitats for diverse biological communities that survive because of these acid environments. When such habitats are treated, the added calcium and neutralized acidity produce conditions that can be toxic to many species present in these important communities. Consequently, management tools for acidic lakes and streams should include not only protocols for neutralizing and reclaiming recently acidified surface waters but also protocols for protecting naturally acidic aquatic ecosystems.

CHAPTER 9

Case Studies

Liming lakes, streams, and watersheds includes many factors, depending on the size, flow, retention, and water quality conditions at the site under consideration. However, the actual liming process is not a complicated procedure, as the following case histories demonstrate. Two of the case histories are examples of a small, one-zone pond and a larger, multi-zone pond. The third demonstrates a watershed liming and the fourth describes a stream liming project. As the examples show, many similarities exist among all the cases.

The basic steps in liming include:

- identifying waterbodies that require liming according to their water quality (pH less than 6.5 and acid neutralizing capacity less than 5 mg/L as $CaCO_3$ [100 µeq/L]);

- developing a treatment plan;

- applying for needed permits;

- informing appropriate state and local authorities and local homeowners;

- contracting with a limestone supplier and treatment contractor; and

- applying the liming material to the site and gauging the timing required for reliming activities.

CASE STUDY 1—LAKE KANACTO

Lake Kanacto is in the Adirondack Mountains of New York state. A small lake (~ 4 ha), it is part of a summer camp that uses it for boating, swimming, and fishing. The fishery in the lake was very poor, and the water quality was not conducive to a healthy fishery (Table 9-1). Strongly stratified, the lake contains a deep, cold, dense water layer at the bottom, a transition layer, and a warm upper layer. Since the lake is small, it was considered as a one-zone lake of 4.05 ha (10 ac).

Table 9-1.—Pre- and post-liming water chemistry and target fisheries summary data for Lake Kanacto, NY (limed August 1986).

YEAR	pH	ANC (μeq/L)	Ca (mg/L)	Al (mg/L)	FISH SPECIES
Pre-liming					
1986	4.92	1.9	0.83	0.032	None captured
Post-liming					
1986	6.61	195.7	4.22	0.039	None captured
1987	6.53	228.0	5.19	0.019	5 Smallmouth bass
1988	6.39	162.3	3.80	0.018	3 Smallmouth bass, 2 Brook trout
1989	6.29	79.8	2.79	0.024	1 Largemouth bass
1990	5.79	38.9	2.06	0.041	1 Largemouth bass

Living Lakes, Inc., a non-profit environmental organization funded by the electric power industry, conducted preliminary tests, which determined that its water quality and the condition of its fishery identified Lake Kanacto as a candidate for liming. Then, Living Lakes compiled a historical record of water quality and biological activity (Simonin et al. 1990) and developed a general treatment plan. It presented these documents to the responsible New York state agencies and the Adirondack Park, all of whom issued permits. Living Lakes also obtained permission from the owners of the summer camp.

Thirty days before treatment, Living Lakes took water samples to determine the specific water quality of the lake and to prescribe how much and what kinds of limestone would best remedy its problems. The test results designated the application of a total of 6.4 tonnes (7.0 tons) of limestone to Lake Kanacto. In the subsequent liming, Living Lakes applied two grades of limestone composed of 95 percent calcium carbonate, 5.4 tonnes (6.0 tons) of one grade to neutralize the water column and 0.9 tonnes (1.0 tons) of a larger grade to provide a sediment dose.

The method of application was to prepare a slurry for a helicopter to spray over the entire surface area of the lake. To prepare the solution and treat the lake, Living Lakes established a staging area at the summer camp close to the lake shore that provided a site for trucks to deliver the limestone as well as a loading and landing area for the helicopter.

As part of the treatment procedure and before the actual liming, Living Lakes sent letters and notices to adjacent landowners, local municipalities, and interested clubs and organizations. Letters to landowners stated the times and days of treatment to alert them to the flyovers by the helicopter; municipalities were alerted to the helicopter flyovers and the traffic from trucks hauling the limestone; and local associations were informed about the environmental effects of the treatment.

At the site, Living Lakes pumped water for the slurry from the lake into the slurry tank on board the helicopter, which was fitted with a Hydro-Spyder sprayer unit and a 300 gallon holding tank. The total carrying capacity of the helicopter was 150 gallons per load because of the high density of the mix. The slurry was a 65 to 70 percent solids suspension, which neutralized the water column over a 72-hour period.

Thirty days after treatment, analysis of water samples showed that it had achieved the acid neutralizing capacity and pH goals for the entire lake. At fall turnover, the limestone in the sediments had mixed with the lake water and achieved the targeted acid neutralizing capacity of greater than 5 mg/L (100 µeq/L) and pH of greater than 6.5 (Table 9-1) and effectively neutralized all strata in the lake.

This result shows that the water column and sediment doses of limestone successfully treated Lake Kanacto for about three years.

CASE STUDY 2—LAWRENCE POND

Lawrence Pond is located in Sandwich, Massachusetts, in the middle of the Cape Cod peninsula. Its surface area covers 55.9 ha (138 ac). Because of the pond's large size and the two deepwater areas it contains, Living Lakes divided Lawrence Pond into four treatment zones. The amount of limestone to be added to each zone was calculated based on area, depth, and measured pH and acid neutralizing capacity. Dry powder limestone was mixed to a 70 percent solids suspension at a staging area near the pond. The total dose was divided among the zones on a volume-weighted average basis to ensure that shallow areas in the pond would receive proportionally less limestone than deeper areas.

By zone, the surface area and slurry dose for treatment were: Zone 1—3.68 ha (9.1 ac), 13.6 tonnes (15.0 tons); Zone 2— 3.9 ha (23.2 ac), 9.3 tonnes (4.3 tons); Zone 3—13.11 ha (32.4 ac), 13.6 tonnes (15.0 tons); and Zone 4—7.41 ha (18.3 ac), 6.1 tonnes (6.7 tons). Since the amount of slurry carried by the helicopter was limited by weight, not volume, the craft flew multiple trips per zone to achieve the required dose.

Lawrence Pond received 34.8 tonnes (38.4 tons) of limestone: as slurried, 49.9 tonnes (55.0 tons), which required 15,000 L (4,000 gal) of water taken from a municipal fire hydrant 15 m (50 ft) from the staging area. A helicopter using a Hydro-Spyder spreader applied the limestone slurry to the site.

Before treatment, the local public was informed of the liming operations. Letters providing information about liming and listing treatment dates went to property owners adjacent to the pond and to chairpersons of local conservation commissions. Notices also were posted at all public access points and staff from the Division of Wildlife and Fisheries were present to ensure that members of the public were not on or in the pond during the liming operation.

Prior to treatment, Living Lakes obtained permits from the Division of Wildlife and Fisheries and the local conservation commission. To acquire these permits, historic as well as pre-liming water quality and biological data that demonstrated the need for liming, such as that listed in Chapter 2, were submitted to the proper authorities. As Table 9-2 shows, the liming successfully achieved a water quality conducive to a healthy fishery. Subsequent stocking of fish proved successful, and the lake was relimed in 1991.

Table 9-2.—Pre- and post-liming water chemistry and target fisheries summary data for Lawrence Pond, MA (limed June 25, 1986).

YEAR	pH	ANC (μeq/L)	Ca (mg/L)	Al (mg/L)	FISH SPECIES
Pre-liming					
1986	6.02	8.02	1.33	0.009	4 Largemouth bass, 1 Smallmouth bass, 1 Chain pickerel
Post-liming					
1986	7.76	201.5	5.45	0.034	14 Largemouth bass
1987	7.09	120.7	4.05	0.012	56 Largemouth bass, 1 Smallmouth bass
1988	7.12	85.8	2.52	0.008	6 Largemouth bass, 4 Smallmouth bass
1989	7.01	78.5	2.66	0.004	1 Largemouth bass, 4 Smallmouth bass
1990	6.53	30.5	1.77	0.008	3 Smallmouth bass

In both case histories, liming proved to be an efficient and economical means of mitigating the effects of acidified surface waters. The acid neutralizing capacity and pH goals were achieved and provided a water quality that was capable of supporting a viable fishery. Recreation, such as swimming and boating, was not inhibited at any time. Negative visual effects of treatment were short-lived; they included a milky, cloudy water color that dissipated within 48 hours of treatment. At each treatment location, public response was overwhelmingly positive.

CASE STUDY 3—WOODS LAKE WATERSHED

Woods Lake is an 8-ha (20-ac) lake in the Adirondack Mountains of New York state. The lake was selected for research on liming in the early 1980s by the Electric Power Research Institute, a nonprofit organization of the electric power industry. The highly acidic lake (pH < 5) was first monitored to establish a baseline of ecosystem behavior. Slurried limestone powder was applied directly to the water surface followed the baseline study. The lake reacidified within 12 to 15 months after application, at which time the liming was repeated.

Because many Adirondack lakes have short hydraulic retention times (median 2 months; Simonin et al. 1990), lake liming may have to be repeated even more frequently to maintain good water quality. Also, nearshore regions and inlet streams may remain acidic when only the lake is limed because acidic waters continue to drain the surrounding watershed.

As a result of the limitations of direct lake liming, EPRI initiated a study to evaluate application of limestone to the soils and wetlands of the surrounding watershed. In October 1989, EPRI limed the watersheds of the two major tributaries of Woods Lake, covering an area of about 35 percent of the total watershed (Fig. 9-1).

Tributary and lake pH, acid neutralizing capacity, and calcium ion all increased after treatment and have remained elevated even during high flows (Porcella et al. 1991). Simultaneously, aluminum concentrations decreased and have remained low. Similar results occurred in surface, organic layers of soil; however, calcite, calcium ions, and acid neutralizing capacity have not penetrated substantially into deeper soil strata of the inorganic layer. About 1 percent of the limestone added to the watershed had dissolved a year after the application. Water quality remained at desirable levels for fish throughout the snowmelt and spring mixing periods. In fact, downstream reaches of the stream draining Woods Lake benefitted for a time after liming. Eventually, these reaches received enough acidic water from adjacent untreated watersheds to reacidify the stream.

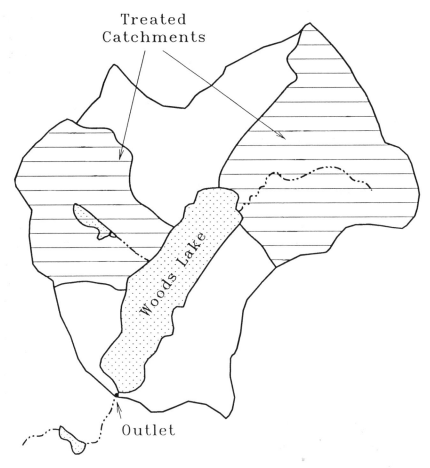

Figure 9-1.—Map of the Woods Lake Watershed. Treated subwatersheds are hatched. The major tributary streams are the north inlet (right) and the west inlet (left), which drains a beaver pond.

Watershed liming also appears to more effectively protect shallower water against acidification during storms compared to direct liming of the lake surface. Fish bioassay results in nearshore areas show that watershed liming has largely eliminated these acidic episodes (Porcella et al. 1991).

In summary, watershed liming restored communities more typical of neutral conditions, provided longer duration of treatment (estimated five or more years versus one to two years with direct lake liming). Moreover, stream and nearshore waters are protected against temporary acidification during storms. Watershed liming is less environmentally intrusive and less costly since fewer incursions into the ecosystem are necessary. Although watershed treatment appears to offer the most complete liming mitigation strategy for acidic lakes and watersheds, long-term assessment will require further measurements of the responses of long-lived species and the duration of desired conditions attained by the treatment.

CASE STUDY 4—DOGWAY FORK

Dogway Fork is a second order stream in the Monongahela National Forest in southeastern West Virginia (Zurbuch et al. 1991). The 13.7-km (8-mile) stream is one of three major tributaries of the Cranberry River. The project is part of a cooperative effort between the U.S. Fish and Wildlife Service and the West Virginia Division of Natural Resources.

Preliming water chemistry data show Dogway Fork to be a highly acidic stream with an average pH of 4.5, alkalinity less than 1.0 mg/L as $CaCO_3$ (< 20 µeq/L), and aluminum as high as 0.6 mg/L. These water chemistry conditions prevent the stream from maintaining a year-round resident fish population.

A self-feeding limestone rotary drum was selected as the limestone dosing device. The treatment system was installed in 1988 at a site 2.9 km (1.7 miles) upstream from the mouth of Dogway Fork. The components include a dam, sluice, two rotary drum assemblies, and a building to store limestone gravel. The station was designed to treat flows up to 3 m^3/sec (100 ft^3/sec) and maintain a pH of 6.5 and an ANC of 2.5 mg/L as $CaCO_3$ (50 µeq/L). The main design features of the system are described on pages 77-79.

During two years of continuous treatment a pH ≥ 6.0 was maintained 91.9 percent of the time. During some high flows the pH fell below 5.0 for short periods of time but calcium levels were maintained above 2.0 mg/L. Total aluminum was reduced from an average of 0.3 mg/L before treatment to a treatment average of 0.26 mg/L. During the two years of treatment a total of 528 tonnes (582 tons) of limestone was used.

No fish were collected in the stream during the first two years of pre-treatment surveys. Fish were observed in Dogway Fork shortly after treatment began and five species were shown to inhabit the stream one year after treatment.

These four case histories provide an overview of an effective means to treat acidified surface waters for achieving water quality that will support an active fisheries population. The Living Lakes Program filmed many lake limings and documented them in a video tape entitled "Upon the Waters." Copies of the tape can be obtained from the Electric Power Research Institute, P.O. Box 10412, Palo Alto, California 94303.

REFERENCES

LITERATURE CITED

Baker, J.P. and S.W. Christensen. 1991. Effects of acidification on biological communities in aquatic ecosystems. Pages 83-106 *in* D.F. Charles, ed. Acidic Deposition and Aquatic Ecosystems: Regional Case Studies. Springer-Verlag, New York.

Binns, N.A. and F.M. Eiserman. 1979. Quantification of fluvial trout habitat in Wyoming. Trans. Am. Fish. Soc. 108:215-28.

Brett, M.T. 1989. Zooplankton communities and acidification processes (a review). Water Air Soil Pollut. 44:387-414.

Brocksen, R.W. and J. Wisniewski, eds. 1988. Restoration of Aquatic and Terrestrial Systems. Kluwer Acad. Publ., Dordredht, The Netherdlands.

Brown, D.J.A. 1982. The effect of pH and calcium on fish and fisheries. Water Air Soil Pollut. 18:343-51.

Brown, J.M. and C.D. Goodyear. 1987. Acid Precipitation Mitigation Program: Research Methods and Protocols. NEC-87/27. U.S. Fish Wildl. Serv. Natl. Ecol. Center, Leetown, WV.

Bukaveckas, P.A. 1989. Effects of calcite treatment on primary producers in acidified Adirondack lakes. II. Short-term response by phytoplankton communities. Can. J. Fish. Aquat. Sci. 46:352-59.

Calhoun, A., ed. 1966. Inland fisheries management. Dep. Fish Game, Resour. Agency, State of Calif., Sacramento.

Carlson, R.E. 1977. A trophic state index for lakes. Limnol. Oceanogr. 22(2):361-68.

Charles, D.F., ed. 1991. Acidic Deposition and Aquatic Ecosystems: Regional Case Studies. Springer-Verlag, New York.

Dennis, W.M. and B.G. Isom, eds. 1984. Ecological Assessment of Macrophyton: Collection, Use, and Meaning of Data. ASTM Spec. Tech. Publ. 843. Am. Soc. Test. Mater., Philadelphia, PA.

DePinto, J.V. et al. 1987. Use of acid lake reacidification model (ALaRM) to assess impact of bottom sediments on calcium carbonate treated lakes. Lake Reserv. Manage. 3:421-29.

Driscoll, C.T., N.M. Johnson, G.E. Likens, and M.C. Feller. 1988. Effects of acidic deposition on the chemistry of headwater streams: a comparison between Hubbard Brook, New Hampshire, and Jamieson Creek, British Columbia. Water Resour. Res. 24(2):195-200.

Driscoll, C.T., G.F. Fordham, W.A. Ayling, and L.M. Oliver. 1989. Short-term changes in the chemistry of trace metals following calcium carbonate treatment of acidic lakes. Can. J. Fish. Aquat. Sci. 46:249-57.

EA Engineering Science and Technology. 1986. Instream Flow Methodologies. Final Rep. EA-4819. Elec. Power Res. Inst., Palo Alto, CA.

Epper, A.W., H.A. Regier, and D.M. Green. 1988. Fish management in New York ponds. Inf. Bull. 116. Coop. Ext. Serv., Cornell Univ., Ithaca, NY.

Espegren, G.D., D.D. Miller, and R.B. Nehring. 1990. Modeling the effects of various angling regulations on trout populations in Colorado streams. Spec. Rep. No. 67. Colo. Div. Wildl., Denver.

Estes, C.C. and I.F. Osborn. 1986. Review and analysis of methods for quantifying instream flow requirements. Water Resour. Bull. 22:389-98.

Evans, R.A. 1989. Response of limnetic insect populations of two acidic, fishless lakes to liming and brook trout (*Salvelinus fontinalis*). Can. J. Fish. Aquat. Sci. 46:342-51.

Everhart, W.H., A.W. Eipper, and W.D. Youngs. 1975. Principles of Fishery Science. Cornell Univ. Press, Ithaca, NY.

Fausch, K.D., C.L. Hawkes, and M.G. Parsons. 1988. Models that predict standing crop of stream fish from habitat variables (1950-1985). Gen. Tech. Rep. PNW-GTR-213. U.S. Dep. Agric. Forest Serv. Pacific Northw. Res. Sta., Portland, OR.

Flick, W.A. and D.W. Webster. 1964. Comparative first year survival and production in wild and domestic strains of brook trout, *Salvelinus fontinalis*. Trans. Am. Fish. Soc. 93:58-69.

Flick, W.A., C.L. Schofield, and D.A. Webster. 1982. Remedial actions for interim maintenance of fish stocks in acidified waters. Pages 287-306 *in* R.E. Johnson, ed. Acid Rain/Fisheries: Proc. Int. Symp. Acidic Precip. Fish. Impacts in Northeast. N. Am. Am. Fish. Soc., Bethesda, MD.

Fraser, J.E. and D.L. Britt. 1982. Liming of Acidified Waters: A Review of Methods and Effects on Aquatic Ecosystems. FWS/OBS-80/40.13. U.S. Fish Wildl. Serv., Kearneysville, WV.

Fraser, J.E. et al. 1985. APMP Guidance Manual—Vol. II: Liming Materials and Methods. Biolog. Rep. 80(40.25). U.S. Fish Wildl. Serv.

Gabelhouse, D.W. Jr., R.L. Hager, and H.E. Klaassen. 1987. Producing Fish and Wildlife from Kansas Ponds. 2nd ed. Kans. Dep. Wildl. Parks, Pratt.

Gloss, S.P., C.L. Schofield, and M.D. Marcus. 1989a. Liming and fisheries management guidelines for acidified lakes in the Adirondack Region. Rep. 80(40.27). U.S. Fish Wildl. Serv. Natl. Ecolog. Res. Center, Leetown, WV.

Gloss, S.P., C.L. Schofield, R.L. Spateholts, and B.A. Plonski. 1989b. Survival, growth, reproduction, and diet of brook trout (*Salvelinus fontinalis*) stocked into lakes after liming to mitigate acidity. Can. J. Fish. Aquat. Sci. 46:277-86.

Gunn, J.M. et al. 1990. Survival, growth, and reproduction of lake trout (*Salvelinus namaycush*) and yellow perch (*Perca flavescens*) after neutralization of an acidic lake near Sudbury, Ontario. Can. J. Fish. Aquat. Sci. 47:446-53.

Hammer, D.A., ed. 1989. Constructed Wetlands for Wastewater Treatment: Municipal, Industrial, and Agricultural. Lewis Publ., Chelsea, MI.

Ivahnenko, T.I., J.J. Rento, and H.W. Rauch. 1988. Effects of liming on water quality of two streams in West Virginia. Water Air Soil Pollut. 41:331-58.

Jackson, M.B. et al. 1990. Effects of neutralization and early reacidification on filamentous algae and macrophytes in Bowland Lake. Can. J. Fish. Aquat. Sci. 47:432-39.

Jansson, M., G. Perrson, and O. Broberg. 1986. Phosphorus in acidified lakes: the example of Lake Gardsjon, Sweden. Hydrobiologia 139:81-96.

Jenkins, R.M. 1982. The morphoedaphic index and reservoir fish production. Trans. Am. Fish. Soc. 111:133-40.

Keller, W., D.P. Dodge, and G.M. Booth. 1990. Experimental lake neutralization program: overview of neutralization studies in Ontario. Can. J. Fish. Aquat. Sci. 47(2):410-11.

Kelly, C.A. et al. 1987. Prediction of biological acid neutralization in acid-sensitive lakes. Biogeochemistry 3:129-40.

Klingbiel, J.H., L.C. Sticker, and O.J. Rongstad. (undated). Wisconsin Farm Fish Ponds. Manual 2. Coop. Ext. Progr., Univ. Wisconsin, Madison.

Kramer, J.R. 1984. Modified Gran analysis for acid and base titrations. Environ. Geochem. Rep. No. 1984-2. McMaster Univ., Hamilton, Ontario.

Lagler, K.F. 1956. Freshwater Fishery Biology. Wm. C. Brown Co., Dubuque, IA.

Lind, O.T. 1985. Handbook of Common Methods in Limnology. Kendall/Hunt Publ. Co., Dubuque, IA.

Marcus, M.D. 1988. Differences in pre- and post-treatment water qualities for twenty limed lakes. Water Air Soil Pollut. 41:279-91.

Marcus, M.D. et al. 1986. An evaluation and compilation of the reported effects of acidification on aquatic biota. Final Rep. EA-4825. Elec. Power Res. Inst., Palo Alto, CA.

Mason, J.W., O.M. Brynidson, and P.E. Degurse. 1967. Comparative survival of wild and domestic strains of brook trout in streams. Trans. Am. Fish. Soc. 96:313-19.

Massachusetts Division of Fisheries and Wildlife. 1984. Statewide Liming of Acidified Waters. Generic Environ. Impact Rep., Mass. Div. Fish. Wildl., Westborough.

McAfee, M.E. 1984. Small Accessible Coldwater Reservoir Studies. Fed. Aid Study F-59-R, Work Plan No. II, Job 1. Rainbow Trout Stocking Evaluations. Colo. Div. Wildl., Fort Collins.

McQuaker, N.R., P.D. Klucker, and D.K. Sandberg. 1983. Chemical analysis of acid precipitation: pH and acidity determinations. Environ. Sci. Tech. 17(7):431-35.

Merritt, R.W. and K.W. Cummins. 1984. An Introduction to the Aquatic Insects of North America. 2nd ed. Kendall/Hunt Publ. Co., Dubuque, IA.

Mills, K.H., S.M. Chalanchuk, L.C. Mohr, and I.J. Davies. 1987. Responses of fish populations in Lake 223 to 8 years of experimental acidification. Can. J. Fish. Aquat. Sci. 44(suppl. 1):114-25.

Molot, L.A., L. Heintsch, and K.H. Nicholls. 1990. Response of phytoplankton in acidic lakes in Ontario to whole-lake neutralization. Can. J. Fish Aquat. Sci. 47:422-31.

Morris, R., E.W. Taylor, and D. Brown, eds. Acid Toxicity and Aquatic Animals. Vol. XXXI, S.E.B. Seminar Ser., Cambridge Univ. Press, Cambridge, England.

Mount, D.R., J.R. Hockett, and W.A. Gern. 1988a. Effect of long-term exposure to acid, aluminum, and low calcium on adult brook trout (*Salvelinus fontinalis*). 2. Vitellogenesis and osmoregulation. Can. J. Fish. Aquat. Sci. 45(9):1633-42.

Mount, D.R. et al. 1988b. Effect of long-term exposure to acid, aluminum, and low calcium on adult brook trout (*Salvelinus fontinalis*). 1. Survival, growth, fecundity, and progeny survival. Can. J. Fish. Aquat. Sci. 45:1623-32.

Mount, D.R. and M.D. Marcus, eds. 1989. Physiologic, Toxicologic, and Population Responses of Brook Trout to Acidification. EPRI EA-6238. Elec. Power Res. Inst., Palo Alto, CA.

Nelson, R.W., G.C. Horak, and J.E. Olson. 1978. Western Reservoir and Stream Habitat Improvements Handbook. FWS/OBS-78/56. Off. Biolog. Serv., Fish Wildl. Serv. U.S. Dep. Inter., Washington, DC.

Nielsen, L.A. and D.L. Johnson. 1983. Fishery Techniques. Am. Fish. Soc., Bethesda, MD.

Nyberg, P., M. Applegate, and E. Degerman. 1986. Effects of liming on crayfish and fish in Sweden. Water Air Soil Pollut. 31:669-87.

Olem, H. 1991. Liming Acidic Surface Waters. Lewis Publ., Chelsea, MI.

Olem, H. et al., eds. 1991. International Lake and Watershed Liming Practices. Terrene Inst., Washington, DC.

Penn Environmental Consultants. 1983. Design Manual: Neutralization of Acid Mine Drainage. EPA-600/2-83-001. U.S. Environ. Prot. Agency, Cincinnati, OH.

Pennak, R.W. 1989. Fresh-water invertebrates of the United States. 3rd ed. John Wiley and Sons, Inc. New York.

Phillips, S.H. 1990. A guide to the construction of freshwater artificial reefs. Sport Fishing Institute, Washington, DC.

Pennsylvania Cooperative Extension Service. 1984. Pennsylvania Fish Ponds. Coll. Agric., Pennsylvania State Univ., University Park.

Plafkin, J.L. et al. 1989. Rapid Bioassessment Protocols for Use in Streams and Rivers. EPA/444/4-89-001. Off. Water, U.S. Environ. Prot. Agency, Washington, DC.

Platts, W.S., W.F. Megahan, and G.W. Minshall. 1983. Methods for evaluating stream, riparian, and biotic conditions. Gen. Tech. Rep. INT-138. U.S. Dep. Agric. Forest Serv., Ogden, UT.

Playe, R.C. and C.M. Wood. 1990. Is precipitation of aluminum fast enough to explain aluminum deposition on fish gills? Can. J. Fish. Aquat. Sci. 47:1558-61.

Porcella, D.B. 1989. Lake acidification mitigation project (LAMP): an overview of an ecosystem perturbation experiment. Can. J. Fish. Aquat. Sci. 46:246-48.

Porcella, D.B. et al. 1991. Limestone treatment for management of aquatic and terrestrial ecosystems. Pages 5-14 *in* H. Olem et al., eds. International Lake and Watershed Liming Practices. Terrene Inst., Washington, DC.

Prescott, G.W. 1970. The Freshwater Algae. W.C. Brown Co. Publ., Dubuque, IA.

Redmond, L.C. 1986. The history and development of warmwater fish harvest regulations. Pages 186-95 *in* G.E. Hall and M.J. Van Den Avyle, eds. Reservoir Fisheries Management: Strategies for the 80's. Reserv. Comm., South. Div. Am. Fish. Soc., Bethesda, MD.

Reiser, D.W., T.A. Wesche, and C. Estes. 1989. Status of instream flow legislation and practices in North America. Fisheries 14:22-29.

Roberts, D.A. and C.W. Boylen. 1989. Effects of liming on the epipelic algal community of Woods Lake, New York. Can. J. Fish. Aquat. Sci. 46:287-94.

Ryder, R.A. 1965. A method for estimating the potential fish production of north-temperate lakes. Trans. Am. Fish. Soc. 94:214-18.

———. 1982. The morphoedaphic index—use, abuse, and fundamental concepts. Trans. Am. Fish. Soc. 111:154-64.

Satterfield, J.R. Jr. and S.A. Flickinger. 1984. Colorado Warmwater Pond Handbook. Fish. Bull. No. 1. Colorado State Univ., Fort Collins.

Scarnecchia, D.L. and E.P Bergersen. 1987. Trout production and standing crop in Colorado's small streams, as related to environmental features. N. Am. J. Fish. Manage. 7:315-30.

Schofield, C.L., S.P. Gloss, B. Plonski, and R. Spateholts. 1989. Production and growth efficiency of brook trout (*Salvelinus fontinalis*) in two Adirondack Mountain (New York) lakes following liming. Can. J. Fish. Aquat. Sci. 46:333-41.

Schrouder, J.D., C.M. Smith, P.J. Rusz, and R.J. White. 1982. Managing Michigan Ponds for Sport Fishing. Ext. Bull. E1554. Coop. Ext. Serv., Michigan State Univ., East Lansing.

Simonin, H.A. et al. 1990. Final Generic Environmental Impact Statement of the New York Department of Environmental Conservation Program of Liming Selected Acidified Waters. New York State Dep. Environ. Conserv., Albany.

Snucins, E.J. 1991. Relative survival of hatchery-reared lake trout, brook trout and F splake stocked in low pH lakes. N. Am. J. Fish. Manage. 11: in press.

Standard Methods for the Examination of Water and Wastewater. 1989. 17th ed. Joint Editorial Board, Am. Pub. Health Assn., Am. Water Works Assn., Water Pollut. Control Fed., Washington, DC.

Stockner, J.G. 1981. Whole lake fertilization for the enhancement of sockeye salmon (*Oncorhynchus nerka*) in British Columbia, Canada. Verh. Int. Verein. Limol. 21:293-99.

Sverdrup, H.U. and P.G. Warfvinge. 1988. What is left for researchers in liming? A critical review of state-of-the-art acidification mitigation. Lake Reserv. Manage. 4:87-97.

U.S. Environmental Protection Agency. 1983. Technical Support Manual: Waterbody Surveys and Assessments for Conducting Use Attainability Analyses. Off. Water Reg. Stand., Washington, DC.

————. 1987. Handbook of Methods for Acid Deposition Studies: Laboratory Analysis for Surface Water Chemistry. EPA 600/4-87/026. Off. Acid Depos., Environ. Monitor. Qual. Assur., Washington, DC.

————. 1989. Assessing Human Health Risks from Chemically Contaminated Fish and Shellfish: A Guidance Manual. EPA 503/8-89-002. Off. Mar. Estuar. Prot., Washington, DC.

Weatherley, N.S. 1988. Liming to mitigate acidification in fresh-water ecosystems: a review of the biological consequences. Water Air Soil Pollut. 39:421-37.

Weber, C.I., ed. 1973. Biological Field and Laboratory Methods for Measuring the Quality of Surface Waters and Effluents. EPA 670/4-73-001. Off. Res. Dev., U.S. Environ. Prot. Agency, Cincinnati, OH.

Weitzel, R.L. 1979. Methods and Measurements of Periphyton Communities: A Review. ASTM Spec. Tech. Publ. 690. Am. Soc. Test. Mater., Philadelphia, PA.

Wesche, T.A. 1985. Stream channel modifications and reclamation structures to enhance fish habitat. Pages 103-63 *in* J.A. Gore, ed. The Restoration of Rivers and Streams, Theories and Experience. Butterworth Publ., Boston, MA.

Wetzel, R.G. 1983. Limnology. 2nd ed. Saunders College Publ., Philadelphia, PA.

Wetzel, R.G. and G.E. Likens. 1990. Limnological Analyses. 2nd ed. Springer-Verlag, New York.

Wood, C.M. et al. 1988a. Physiological evidence of acclimation to acid/aluminum stress in adult brook trout (*Salvelinus fontinalis*). 1. Blood composition and net sodium fluxes. Can. J. Fish. Aquat. Sci. 45:1587-96.

Wood, C.M., B.P. Simons, D.R. Mount, and H.L. Bergman. 1988b. Physiological evidence of acclimation to acid/aluminum stress in adult brook trout (*Salvelinus fontinalis*). 2. Blood parameters by cannulation. Can. J. Fish. Aquat. Sci. 45:1597-1605.

Zurbuch, P.E., R. Menendez, and J.L. Clayton. 1991. West Virginia Dogway Fork Project Cooperative Mitigation Program. Progr. Rep. 1990. U.S. Fish Wildl. Serv., Kearneysville, WV.

ADDITIONAL SOURCES

Annear, T.C. and A.L. Conder. 1984. Relative bias of several fisheries instream flow methods. N. Am. J. Fish. Manage. 4:531-39.

Arce, R.G. and D.E. Boyd. 1975. Effects of agricultural limestone on water chemistry, phytoplankton productivity, and fish production in soft water ponds. Trans. Am. Fish. Soc. 104(2):308-12.

Bergman, H.L., J.S. Mattice, and D.J.A. Brown. 1988. Lake acidification and fisheries project: adult brook trout. Can. J. Fish. Aquat. Sci. 45:1561-62

Brown, D.J.A. 1988. The Loch Fleet and other catchment liming programs. Water Air Soil Pollut. 41:409-16.

Buckman, H.D. and N.C. Brady. 1966. The Nature and Properties of Soils. 6th ed. MacMilliam Co., New York.

Elser, M.M., J.J. Elser, and S.R. Carpenter. 1986. Peter and Paul Lakes: a liming experiment revisited. Am. Midland Nat. 116(2):282-95.

Gloss, S.P., C.L. Schofield, and R.L. Spateholts. 1987. Conditions for reestablishment of brook trout (*Salvelinus fontinalis*) populations in acidic lakes following base addition. Lake Reserv. Manage. (3):412-20.

Green, R.H. 1979. Sampling Design and Statistical Methods for Environmental Biologists. John Wiley and Sons, New York.

Hayes, M.L. 1983. Active fish capture methods. Pages 123-45 *in* L.A. Nielsen and D.L. Johnson, eds. Fisheries Techniques. Am. Fish. Soc., Bethesda, MD.

Hubert, W.A. 1983. Passive capture techniques. Pages 95-122 *in* L.A. Nielsen and D.L. Johnson, eds. Fisheries Techniques. Am. Fish. Soc., Bethesda, MD.

Kieth, W.E. 1986. A review of introduction and maintenance stocking in reservoir fisheries management. Pages 144-48 *in* G.E. Hall and M.J. Van Den Avyle, eds. Reservoir Fisheries Management: Strategies for the 80's. Reserv. Comm., South. Div. Am. Fish. Soc., Bethesda, MD.

Keller, W., L.A. Molot, R.W. Griffiths, and N.D. Yan. 1990. Changes in the zoobenthos community of acidified Bowland Lake after whole-lake neutralization and lake trout (*Salvelinus namaycush*) reintroduction. Can. J. Fish. Aquat. Sci. 47:440-45.

Malvestuto, S.P. 1983. Sampling the recreational fishery. Pages 397-419 *in* L.A. Nielsen and D.L. Johnson, eds. Fisheries Techniques. Am. Fish. Soc., Bethesda, MD.

Marcus, M.D., M.K. Young, L.E. Noel, and B.A. Mullan. 1990. Salmonid-habitat Relationships in the Western United States: A Review and Indexed Bibliography. Gen. Tech. Rep. RM-188. U.S. Dep. Agric. Forest Serv. Rocky Mount. Forest Range Exp. Sta., Fort Collins, CO.

McDonald, D.G., J.P. Reader, and T.R.K. Dalziel. 1989. The combined effects of pH and trace metals on fish inoregulation. Pages 221-42 *in* R. Morris, E.W. Taylor, D.J.A. Brown, and J.A. Brown, eds. Acid Toxicity and Aquatic Animals. Cambridge Univ. Press.

Oglesby, R.T. 1982. The MEI Symposium—overview and observations. Trans. Am. Fish. Soc. 111:171-75.

Ricker, W.E. 1975. Computation and interpretation of biological statistics of fish populations. Bull. Fish. Res. Board Can. 191.

Schindler, D.W. et al. 1985. Long-term ecosystem stress: the effects of years of experimental acidification on a small lake. Science 228:1395-401.

Vollenweider, R.A. A Manual on Methods for Measuring Primary Production in Aquatic Environments. Blackwell Sci. Publ., Oxford and Edinburgh, England.

APPENDIX

Sample Forms for Limestone Application and Monitoring

FORM LT1

TREATMENT INFORMATION

SITE NAME	SITE ID	TREATMENT OPERATION

ANTICIPATED TREATMENT DATE	ALTERNATE TREATMENT DATE (1)	ALTERNATE TREATMENT DATE (2)	FIELD SUPER. ID
_ _ _ _ _ _ D D M M M Y Y	_ _ _ _ _ _ D D M M M Y Y	_ _ _ _ _ _ D D M M M Y Y	_ _ _

ACTUAL START DATE	ACTUAL END DATE	TREATMENT TIME
_ _ _ _ _ _	_ _ _ _ _ _	_ _ _ hrs

WATER QUALITY OBJECTIVES

pH _ . _   ANC _ _ _ µeq/l   SEDIMENT DOSE _ _ . _ TONNES CaCO$_3$ /ha

TREATMENT TECHNOLOGY

APPLICATION TECHNOLOGY
☐ Helicopter
☐ Barge
☐ Other _____

MATERIAL AS DELIVERED ☐ Dry ☐ Slurry
MATERIAL AS APPLIED ☐ Dry ☐ Slurry
DISPERSANT REQUIRED? ☐ YES ☐ NO

WATER SUPPLY
☐ Not required
☐ Site
☐ Municipal
☐ Other _____

STAGING AREA   NAME

Distance from site _ _ . _ km

N. LATITUDE	W. LONGITUDE
_ _ . _ _ _ D D M M M	_ _ _ . _ _ _ D D D M M M

SITE PREPARATION _____

TREATMENT ZONE MORPHOMETRY

Zone No.	Mean Depth (m)	Surface Area (ha)	Zone No.	Mean Depth (m)	Surface Area (ha)

DESIGN DOSE

Zone No.	Material ID	Dry Weight (Tonnes)	Zone No.	Material ID	Dry Weight (Tonnes)

	MATERIAL ID	DATE OF ANALYSIS
FORM LT2 **MATERIAL IDENTIFICATION** **CHEMICAL PROPERTIES**	— — — — ☐ ☐	$\overline{D}$ $\overline{D}$ $\overline{M}$ $\overline{M}$ $\overline{Y}$ $\overline{Y}$

PARAMETER	TECHNICAL DATA SHEET VALUE	ANALYTICAL LABORATORY VALUE
$CaCO_3$ (Wt. %)	— — . — —	— — . — —
$MgCO_3$ (Wt. %)	— — . — —	— — . — —
Neut.Cap.$CaCO_3$ (Wt. %)	— — . — —	— — . — —
pH	— — . — —	— — . — —
Tot. Org.C (mg/l)	— — . — —	— — . — —
P (Wt. %)	— — . — —	— — . — —
NO_3 (mg/kg)	— — . — —	— — . — —
NH_3 (mg/kg)	— — . — —	— — . — —
Loss @ 200° C (Wt. %)	— — . — —	— — . — —
Acid Insol. Mat. (Wt. %)	— — . — —	— — . — —
Al (Wt. %)	— — . — —	— — . — —
Mg & Alkali Salts (Wt. %)	— — . — —	— — . — —
Pb (ppm)	— — . —	— — . —
As (ppm)	— — . —	— — . —
Fl (ppm)	— — . —	— — . —
Mg (ppm)	— — . —	— — . —
Cd (ppm)	— — . —	— — . —
Cr (ppm)	— — . —	— — . —
Co (ppm)	— — . —	— — . —
Cu (ppm)	— — . —	— — . —
Mn (ppm)	— — . —	— — . —
Hg (ppm)	— — . —	— — . —
Ni (ppm)	— — . —	— — . —
V (ppm)	— — . —	— — . —
Zn (ppm)	— — . —	— — . —

10073/CHEMICAL PROPERTIES/C002

167

MATERIAL INFORMATION PHYSICAL PROPERTIES FORM LT3	MATERIAL ID — — — —	DETERMINATION DATE D̅ D̅ M̅ M̅ Y̅ Y̅

PARTICLE DIAMETER (μm)	WT% PASSING
250	— —
125	— —
64	— —
32	— —
16	— —
8	— —
4	— —
2	— —
1	— —

DETERMINATION METHOD

(check one)

☐ AIRSIEVE   ☐ HYDROMETER

☐ ANDRASSON PIPETTE   ☐ LASER SEDIGRAPH

☐ CENTRIFUGE SEDIGRAPH   ☐ SEDIGRAPH

☐ COULTER COUNTER   ☐ WET SEIVING

10073/Material Info Phys/C002

		TREATMENT OPERATION	STAGING AREA

**FIELD
BATCH INFORMATION
FORM LT4**

			DATE	FIELD SUPERVISOR ID	FIELD SUPERVISOR'S INITIALS
			D D M M Y Y	— — —	

BATCH ID	BATCH VOLUME (GALLONS)	SOLIDS CONTENT (%)	SAMPLE ID	VENDOR SAMPLE ID
— — —	— — —	— — . —	— — —	_____
— — —	— — —	— — . —	— — —	_____
— — —	— — —	— — . —	— — —	_____
— — —	— — —	— — . —	— — —	_____
— — —	— — —	— — . —	— — —	_____
— — —	— — —	— — . —	— — —	_____
— — —	— — —	— — . —	— — —	_____
— — —	— — —	— — . —	— — —	_____
— — —	— — —	— — . —	— — —	_____
— — —	— — —	— — . —	— — —	_____
— — —	— — —	— — . —	— — —	_____
— — —	— — —	— — . —	— — —	_____
— — —	— — —	— — . —	— — —	_____
— — —	— — —	— — . —	— — —	_____
— — —	— — —	— — . —	— — —	_____
— — —	— — —	— — . —	— — —	_____
— — —	— — —	— — . —	— — —	_____
— — —	— — —	— — . —	— — —	_____
— — —	— — —	— — . —	— — —	_____
— — —	— — —	— — . —	— — —	_____

10073/Field Batch Info/C002

		TREATMENT OPERATION	STAGING AREA

FIELD BATCH TRUCK FORM
FORM LT5

DATE	FIELD SUPERVISOR ID	FIELD SUPERVISOR'S INITIALS
D D M M M Y Y _ _ _ _ _ _	_ _ _	

TRUCK BILL OF LADING NO.	MATERIAL ID	BATCH ID	MATERIAL TRANSFERRED (DRY TONS)	DRY SAMPLE ID
_____	_ _ _ _	_ _ _ _	_ _ . _	_ _ _ _
_____	_ _ _ _	_ _ _ _	_ _ . _	_ _ _ _
_____	_ _ _ _	_ _ _ _	_ _ . _	_ _ _ _
_____	_ _ _ _	_ _ _ _	_ _ . _	_ _ _ _
_____	_ _ _ _	_ _ _ _	_ _ . _	_ _ _ _
_____	_ _ _ _	_ _ _ _	_ _ . _	_ _ _ _
_____	_ _ _ _	_ _ _ _	_ _ . _	_ _ _ _
_____	_ _ _ _	_ _ _ _	_ _ . _	_ _ _ _
_____	_ _ _ _	_ _ _ _	_ _ . _	_ _ _ _
_____	_ _ _ _	_ _ _ _	_ _ . _	_ _ _ _
_____	_ _ _ _	_ _ _ _	_ _ . _	_ _ _ _
_____	_ _ _ _	_ _ _ _	_ _ . _	_ _ _ _
_____	_ _ _ _	_ _ _ _	_ _ . _	_ _ _ _
_____	_ _ _ _	_ _ _ _	_ _ . _	_ _ _ _
_____	_ _ _ _	_ _ _ _	_ _ . _	_ _ _ _
_____	_ _ _ _	_ _ _ _	_ _ . _	_ _ _ _
_____	_ _ _ _	_ _ _ _	_ _ . _	_ _ _ _
_____	_ _ _ _	_ _ _ _	_ _ . _	_ _ _ _
_____	_ _ _ _	_ _ _ _	_ _ . _	_ _ _ _
_____	_ _ _ _	_ _ _ _	_ _ . _	_ _ _ _

10073/Field Batch Truck/C002

		TREATMENT OPERATION	STAGING AREA	

FIELD BATCH ZONE FORM
FORM LT6

DATE	FIELD SUPERVISOR ID	FIELD SUPERVISOR'S INITIALS
‾D‾ ‾D‾ ‾M‾ ‾M‾ ‾Y‾ ‾Y‾	— — —	

BATCH ID	SITE ID	ZONE NO.	MATERIAL TRANSFERRED (DRY TONS)
— — — —	— — — — — —	— —	— — · — —
— — — —	— — — — — —	— —	— — · — —
— — — —	— — — — — —	— —	— — · — —
— — — —	— — — — — —	— —	— — · — —
— — — —	— — — — — —	— —	— — · — —
— — — —	— — — — — —	— —	— — · — —
— — — —	— — — — — —	— —	— — · — —
— — — —	— — — — — —	— —	— — · — —
— — — —	— — — — — —	— —	— — · — —
— — — —	— — — — — —	— —	— — · — —
— — — —	— — — — — —	— —	— — · — —
— — — —	— — — — — —	— —	— — · — —
— — — —	— — — — — —	— —	— — · — —
— — — —	— — — — — —	— —	— — · — —
— — — —	— — — — — —	— —	— — · — —
— — — —	— — — — — —	— —	— — · — —
— — — —	— — — — — —	— —	— — · — —
— — — —	— — — — — —	— —	— — · — —
— — — —	— — — — — —	— —	— — · — —
— — — —	— — — — — —	— —	— — · — —

10073/Field Batch Zone/C002

				TREATMENT OPERATION	
FORM LT7 **FIELD BULK TRUCK LOG**				_____	
MATERIAL CODE	BILL OF LADING #	VENDOR SAMPLE ID	SAMPLE ID	NET BULK WEIGHT (TONS)	DATE/TIME DELIVERED
— — —			— — — —	— — . —	— — — — — — — — — —
— — —			— — — —	— — . —	— — — — — — — — — —
— — —			— — — —	— — . —	— — — — — — — — — —
— — —			— — — —	— — . —	— — — — — — — — — —
— — —			— — — —	— — . —	— — — — — — — — — —
— — —			— — — —	— — . —	— — — — — — — — — —
— — —			— — — —	— — . —	— — — — — — — — — —
— — —			— — — —	— — . —	— — — — — — — — — —
— — —			— — — —	— — . —	— — — — — — — — — —
— — —			— — — —	— — . —	— — — — — — — — — —
— — —			— — — —	— — . —	— — — — — — — — — —
— — —			— — — —	— — . —	— — — — — — — — — —
— — —			— — — —	— — . —	— — — — — — — — — —
— — —			— — — —	— — . —	— — — — — — — — — —
— — —			— — — —	— — . —	— — — — — — — — — —
— — —			— — — —	— — . —	— — — — — — — — — —
— — —			— — — —	— — . —	— — — — — — — — — —
— — —			— — — —	— — . —	— — — — — — — — — —
— — —			— — — —	— — . —	— — — — — — — — — —
— — —			— — — —	— — . —	— — — — — — — — — —
— — —			— — — —	— — . —	— — — — — — — — — —
— — —			— — — —	— — . —	— — — — — — — — — —

10073/FIELD BULK TRUCK/C002

MATERIAL DESCRIPTION
FORM LT8

Material I.D. ___ ___ ___ ___

Vendor_____

Material Trade Name_____

Material Origin _____
(Quarry, City, State)

10073/Material Description/C002

173

GLOSSARY

Absence — when used in the context of the presence or absence of a fish species or other organisms, species not caught during sampling (generally using a standard sampling regime for all waters sampled) are defined as absent.

Abundance — the number of organisms per unit area or volume.

Acid mine drainage — runoff with high concentrations of metals and sulfate and high levels of acidity resulting from the oxidation of sulfide minerals that have been exposed to air and water by mining activities.

Acid neutralizing capacity (ANC) — the equivalent capacity of a solution to neutralize strong acids. The components of ANC include weak bases (carbonate species, dissociated organic acids, alumino-hydroxides, borates, and silicates) and strong bases (primarily, OH^-). ANC often is measured by the Gran titration procedure.

Acidic lake or stream — a lake or stream in which the acid neutralizing capacity is less than or equal to 0.

Acidification — the decrease of *acid neutralizing capacity* in water or *base saturation* in soil caused by natural or anthropogenic processes; includes both decreases in surface water pH and all the associated water chemistry changes common to these waters.

Acidified — pertaining to a natural water that has experienced a decrease in acid neutralizing capacity or a soil that has experienced a reduction in base saturation.

Acidophilic — "acid loving," that is, describing organisms that thrive in an acidic environment.

Adaptation — a change in the sensitivity of an organism to an environmental stress or factor over time.

Alkalinity — for this report, the equivalent sum of $HCO_3^- + CO_3^{2-} + OH^-$ minus H^+, i.e., buffering conferred by the bicarbonate system; the terms ANC and alkalinity are sometimes used interchangeably. ANC includes alkalinity plus additional buffering from dissociated organic acids and other compounds. See *acid neutralizing capacity*.

Anion — a negatively charged ion.

Battery-powered doser — small stream liming device typically powered by a 12-volt battery.

Bedrock — solid rock exposed at the surface of the earth or overlain by saprolites or unconsolidated material.

Benthic invertebrates — insects or other organisms lacking a spinal column that live in association with the lake or stream bottom.

Benthic — bottom zones or bottom-dwelling organisms in water bodies.

Biological community — a general collective term to describe the various species of organisms (multiple species) living together in a given ecosystem.

Biological recovery — an observed (or projected) sustained improvement in indicators of biological condition at the species, population, or community level, in response to actual (or simulated) decreased acidic conditions.

Biomass — the total quantity of living biological material in units of weight or mass, usually within a volume or area of water.

Calibration — process of checking, adjusting, or standardizing operating characteristics of instruments and model appurtenances on a physical model or coefficients in a mathematical model with empirical data of known quality. The process of evaluating the scale readings of an instrument with a known standard in terms of the physical quantity to be measured.

Catchment — see *watershed*.

Cation — a positively charged ion.

Circumneutral — close to neutrality with respect to pH (neutral pH = 7); in natural waters, pH 6-8.

Color — hues present in water itself, resulting from colloidal and dissolved substance; generally measured by comparing water to standard solutions of platinum-cobalt dyes or colored glass disks.

Conductance (or conductivity) — the ability of a substance (e.g., aqueous solution) to carry an electrical current. For natural waters, conductance is closely related to the total concentration of dissolved ions. Conductance is usually measured under very specific conditions.

Community — see *biological community*.

Decomposition — the microbially mediated reaction that converts solid or dissolved organic matter into its constituents (also called decay or mineralization).

Density — mass per unit volume; also see *abundance*.

Detritus — dead and decaying organic matter originating from plants and animals.

Discharge areas — geographic portions of a watershed where there is surface flow, particularly during periods of stormflow or snowmelt.

Dispersant — chemical agent added to limestone to improve dissolution and minimize settling.

Dissolution efficiency — ratio between the dissolved amount of neutralizing material in water and the total added amount over a given unit of time.

Dissolved inorganic carbon — the sum of dissolved carbonic acid, carbon dioxide, bicarbonate, and carbonate in a water sample.

Dissolved organic carbon — organic carbon that is dissolved or unfilterable in a water sample using generally 0.45 μm pore size filters.

Diversion well — a cylindrical concrete structure containing crushed limestone through which stream water is diverted to allow mechanical grinding of limestone and subsequent neutralization.

Dose — the quantity of material added per unit volume or per unit of time.

Doser — any mechanical device designed to continuously treat acidic flowing waters by the addition of base materials.

Dry-powder doser — an automated device that stores dry limestone powder and dispenses either dry powder or slurried powder.

Dystrophic — surface waters low in nutrients yet highly colored with dissolved humic organic matter.

Fingerlings — post-larval juvenile fish; "a small fish no longer than a finger."

Flushing rate — the reciprocal of retention.

Gran analysis — a mathematical procedure used to determine the equivalence points of a titration curve for acid neutralizing capacity.

Groundwater — water in a saturated zone within soil or rock.

Hydrology — the science that treats the waters of the earth their occurrence, circulation, and distribution; their chemical and physical properties; and their reaction with their environment, including their relationship to living things.

Limestone barrier — barrier of limestone aggregate placed across a stream channel to neutralize acid water.

Liming — the addition of any base materials to neutralize surface water or sediment or to increase acid neutralizing capacity.

Littoral zone — the shallow, nearshore region of a body of water; often defined as the band from the shoreline to the outer edge of the occurrence of rooted vegetation.

Macrophytes — large forms of aquatic vegetation that can be seen easily with the eye.

Mineral weathering — dissolution of rocks and minerals by chemical and physical processes.

Mitigation — generally described as amelioration of adverse impacts at the source (e.g., discharge reductions) or the receptor (e.g., lake liming). For the purposes of this report, amelioration of acidic conditions in surface waters to preserve or restore fisheries and the supporting aquatic community.

Nutrient cycling — the movement or transfer of chemicals required for biological maintenance or growth among components of the ecosystem by physical, chemical, or biological processes.

Oligotrophic lake — describes a lake with little organic matter and little biological activity.

Organic acids — acids possessing a carboxyl (-COOH) group or phenolic (C-OH) group; includes fulvic and humic acids.

Periphyton — the communities of microscopic aquatic organisms colonizing submerged surfaces; most analyses of this community emphasize the algae populations.

pH — the negative logarithm of the hydrogen ion activity. The pH scale extends from 1 (most acidic) to 14 (most alkaline); a difference of one pH unit indicates a tenfold change in hydrogen ion activity.

Phytoplankton — the community of microscopic, free-floating algal cells inhabiting water columns in ponds, lakes, reservoirs, and larger rivers.

Platinum-cobalt units — a standard measure of the true color in water; the color produced by 1.0 mg of platinum in 1.0 L of water.

Pool — (1) in streams, a relatively deep area with low velocity; (2) in ecological systems, the supply of an element or compound, such as exchangeable or weatherable cations or adsorbed sulfate, in a defined component of the ecosystem.

Precision — a measure of the capacity of a method to provide reproducible measurements of a particular analyte (often represented by variance).

Predict — to estimate some current or future condition within specified confidence limits on the basis of analytical procedures and historical or current observations.

Prediction — a calculation or estimation of some current or future condition on the basis of historical or current observation, experience, or scientific reason. (See *predict*.)

Quality assurance — a system of activities for which the purpose is to provide assurance that a product (e.g., database) meets a defined standard of quality with a stated level of confidence.

Quality control — steps taken during sample collection and analysis to ensure that data quality meets the minimum standards established in a quality assurance plan.

Reacidification — see *acidification*.

Recovery — see *biological recovery*.

Respiration — the biological oxidation of organic carbon with concomitant reduction of external oxidant and the production of energy.

Retention time — the estimated mean time (usually expressed in years) that water remains in a lake before it leaves the system.

Riffle — in streams, a relatively shallow area with high velocity water flow extending across a stream in which water flow is broken at the surface.

Rotary drum — water-powered liming device for streams in which limestone aggregate is ground in a revolving drum.

Secchi disk depth — a measure of the transparency of water by use of a round disk that is lowered into a waterbody until it can no longer be seen.

Silo — a storage bin, in this case to store limestone.

Slurry — a liquid mixture containing suspended limestone particles.

Species richness — the number of species occurring in a given aquatic ecosystem, generally estimated by the number of species caught when a standard sampling technique is used.

Staff gauge — a measuring device used to quantify stream stage (depth); usually made from a narrow strip of sheet metal and attached to a vertical post or wall of stream flume and calibrated to provide measures of streamflow volumes from readings of stream stage.

Stream stage — relative depth of the stream, generally quantified using a staff gauge.

Strong acids — acids with a high tendency to donate protons or to completely dissociate in natural waters, e.g., H_2SO_4, HNO_3, HCl, and some organic acids.

Surface area-weighted mean particle diameter — average diameter weighted to an individual particle's surface area.

Tonne — metric ton or 1000 kg; convert to tons by multiplying by 0.9072.

Turbidity — the cloudiness of water caused by suspended particles.

Turnover — (1) the tendency for the waters of lakes to mix vertically when the surface layers warm in the spring or cool in the fall to 4°C (39°F), the temperature where water is most dense, and sink toward the bottom; (2) when used as turnover rate, a synonym for lake or reservoir flushing rate.

$\mu eq/L$ — 10^{-6} equivalents per liter; one gram equivalent weight of an acid is the quantity of acid that can donate 1 mole of protons (H^+) to a base.

Variable — a quantity that may assume any one of a set of values during analysis.

Water-powered doser — mechanical device that uses the energy of running water to operate a mechanism to dispense limestone powder

Watershed — the geographic area from which surface water drains into a particular lake or point along a stream.

Watershed liming — neutralization of surface waters by applying limestone to the terrestrial portion of the watershed. The areas may include wetlands, recharge areas (dry soils), and discharge areas.

Wetland — area that is inundated or saturated by surface or groundwater at a frequency and duration sufficient to support a prevalence of vegetation typically adapted for life in saturated soil.

Zooplankton — the community of microscopic animals, usually with weak swimming abilities that inhabit water columns in ponds, lakes, reservoirs, and larger rivers.

INDEX

Trophy fisheries, 35, 98
Trout, 119-120. *See also* specific
  type
Trucks, 56
  for transporting limestone
    materials, 51-52
  in lake liming, 60, 63-64
  in watershed liming, 82
Turbidity, 180
  liming effects on, 130, 149
Turnover, 147, 180
  lake liming during, 57
  timing of monitoring, 109

Underpass deflectors, 104
Undesirable fish species, 87-89,
  96. *See also* Target fish species
U.S. Fish and Wildlife Service, 2

Vegetation, aquatic, 41. *See also*
  Macrophytes
Velocity meter, 121
Volume flow, streams, 77,
  104-106, 120-122
Volume, lake, 19-20, 41, 58, 61,
  122-123

Walleye, 28, 33-35
Warmouth, 34
Warmwater fish, 32-33, 35. *See
  also* specific type
Waste organics, 17
Water chemistry, 4, 96
  analysis, 112, 114
  for Lake Kanacto, 146
  for Lawrence Pond, 149
Water-powered dosers, 71, 74,
  180
  costs of, 24, 80
  for stream liming, 74, 80
Water quality
  acidification effects on, 7

data for permit application, 40
goals, 25-29, 41, 107
limitations, 14-17
management, 85-87
monitoring of changes in, 41,
  107-108, 110
Water retention time, 19, 41,
  54-57, 61-62, 145, 150
Water samples analysis, 112,
  114-115
Water temperature
  fish tolerance for, 32-33, 35
  liming dosage and, 58, 69
  measurement of, 116-117
  rotenone and, 87
Watershed liming, 4, 54-55,
  79-83, 149-152, 180
  boats in, 82
  costs of, 22, 83
  dosage calculations for, 82
  fixed-wing aircraft in, 83
  helicopters in, 83
  location of, 81-82
  timing of, 81
  trucks and tractors in, 82
Weirs, 104
West Virginia, permit process, 40
  rotary drum system, 79
Wetland liming, 3, 54, 150
White crappie, 34, 120
White sucker, 9, 28
Winter fishkills, 17, 20, 98
Woods Lake watershed, 149-152

Year class structure, of fishery,
  112
Yellow perch, 28, 33-36

Zones. *See* Multi-zone lake; One
  zone lake
Zooplankton, liming effects on,
  136-137